Intelligent Agents and Their Applications

Studies in Fuzziness and Soft Computing

Editor-in-chief
Prof. Janusz Kacprzyk
Systems Research Institute
Polish Academy of Sciences
ul. Newelska 6
01-447 Warsaw, Poland
E-mail: kacprzyk@ibspan.waw.pl
http://www.springer.de/cgi-bin/search_book.pl?series=2941

Further volumes of this series can be found at our homepage.

Vol. 79. A. Osyczka
Evolutionary Algorithms for Single and Multicriteria Design Optimization, 2002
ISBN 3-7908-1418-0

Vol. 80. P. Wong, F. Aminzadeh and M. Nikravesh (Eds.)
Soft Computing for Reservoir Characterization and Modeling, 2002
ISBN 3-7908-1421-0

Vol. 81. V. Dimitrov and V. Korotkich (Eds.)
Fuzzy Logic, 2002
ISBN 3-7908-1425-3

Vol. 82. Ch. Carlsson and R. Fullér
Fuzzy Reasoning in Decision Making and Optimization, 2002
ISBN 3-7908-1428-8

Vol. 83. S. Barro and R. Marín (Eds.)
Fuzzy Logic in Medicine, 2002
ISBN 3-7908-1429-6

Vol. 84. L.C. Jain and J. Kacprzyk (Eds.)
New Learning Paradigms in Soft Computing, 2002
ISBN 3-7908-1436-9

Vol. 85. D. Rutkowska
Neuro-Fuzzy Architectures and Hybrid Learning, 2002
ISBN 3-7908-1438-5

Vol. 86. M.B. Gorzałczany
Computational Intelligence Systems and Applications, 2002
ISBN 3-7908-1439-3

Vol. 87. C. Bertoluzza, M.Á. Gil and D.A. Ralescu (Eds.)
Statistical Modeling, Analysis and Management of Fuzzy Data, 2002
ISBN 3-7908-1440-7

Vol. 88. R.P. Srivastava and T.J. Mock (Eds.)
Belief Functions in Business Decisions, 2002
ISBN 3-7908-1451-2

Vol. 89. B. Bouchon-Meunier, J. Gutiérrez-Ríos, L. Magdalena and R.R. Yager (Eds.)
Technologies for Constructing Intelligent Systems 1, 2002
ISBN 3-7908-1454-7

Vol. 90. B. Bouchon-Meunier, J. Gutiérrez-Ríos, L. Magdalena and R.R. Yager (Eds.)
Technologies for Constructing Intelligent Systems 2, 2002
ISBN 3-7908-1455-5

Vol. 91. J.J. Buckley, E. Eslami and T. Feuring
Fuzzy Mathematics in Economics and Engineering, 2002
ISBN 3-7908-1456-3

Vol. 92. P.P. Angelov
Evolving Rule-Based Models, 2002
ISBN 3-7908-1457-1

Vol. 93. V.V. Cross and T.A. Sudkamp
Similarity and Compatibility in Fuzzy Set Theory, 2002
ISBN 3-7908-1458-X

Vol. 94. M. MacCrimmon and P. Tillers (Eds.)
The Dynamics of Judical Proof, 2002
ISBN 3-7908-1459-8

Vol. 95. T.Y. Lin, Y.Y. Yao and L.A. Zadeh (Eds.)
Data Mining, Rough Sets and Granular Computing, 2002
ISBN 3-7908-1461-X

Vol. 96. M. Schmitt, H.-N. Teodorescu, A. Jain, A. Jain, S. Jain and L.C. Jain (Eds.)
Computational Intelligence Processing in Medical Diagnosis, 2002
ISBN 3-7908-1463-6

Vol. 97. T. Calvo, G. Mayor and R. Mesiar (Eds.)
Aggregation Operators, 2002
ISBN 3-7908-1468-7

Lakhmi C. Jain · Zhengxin Chen
Nikhil Ichalkaranje
Editors

Intelligent Agents and Their Applications

With 119 Figures
and 10 Tables

Springer-Verlag Berlin Heidelberg GmbH

Professor Lakhmi C. Jain
University of South Australia
Knowledge-Based Intelligent
Engineering Systems Centre
Mawson Lakes, Adelaide
South Australia 5095
Lakhmi.Jain@unisa.edu.au

Professor Zhengxin Chen
University of Nebraska
Department of Computer Science
Omaha, NE 68182-0500
USA
zchen@mail.unomaha.edu

Nikhil Ichalkaranje
University of South Australia
Knowledge-Based Intelligent
Engineering Systems Centre
Mawson Lakes, Adelaide
South Australia 5095

ISSN 1434-9922

ISBN 978-3-7908-2510-7 ISBN 978-3-7908-1786-7 (eBook)
DOI 10.1007/978-3-7908-1786-7

Cataloging-in-Publication Data applied for
Die Deutsche Bibliothek – CIP-Einheitsaufnahme
Intelligent agents and their applications: with 10 tables / Lakhmi C. Jain ... (ed.). – Heidelberg; New York: Physica-Verl., 2002
(Studies in fuzziness and soft computing; Vol. 98)

Hardcover Design: Erich Kirchner, Heidelberg

DEDICATION

This book is dedicated to our students.

The proceeds of royalty will be donated to a Charity.

Foreword

An agent is normally defined as a computer program that continuously and autonomously acts on behalf of the user. In practice, varieties of artifacts, such as synthetic lifelike conversational characters on the computer display, computer programs that broker information on the information network, or computing units that collectively exhibit an intelligent behavior, are called agents.

Why are people so excited about agents? The current enthusiasm about the agent technology may be traced back to the beginning of the 1990s, five years after Minsky's "The Society of Mind" was published, a few years after Apple's Knowledge Navigator promotion video was released to the public, and four years prior to Communications of the ACM's special issue on intelligent agents (Volume 37, No. 7, July 1994). More than ten years after those "good old days," the enthusiasm does not appear to cease, but on the contrary is becoming even widespread in society. In the beginning, strong suspicion coexisted with the enthusiasm. Critics thought the agent technology was no more than a fad that would soon disappear. After a while, it became mode and fashion, and today it is a well-recognized standard branch of the information and communication technology.

Agent technology entertains researchers with thousands of challenging goals. Compared with those of ten years ago, researchers now address more practical goals ranging from knowledge management and supply chains to collective robotics. These real world problems urge them to solve real, hard issues that are not only significant in practical applications but also truly interesting from the theoretical points of view. Today's agent technology stands on a large amount of innovations driven by real world applications and evaluations. Practitioners are happy with agents, too. Agent technology presents a huge potential for business. In fact, agent technology has been successfully applied to numerous business applications ranging from entertainment and education to electronic commerce and reliable intelligent manufacturing.

Indeed, a closer look at agent technology reveals that it is a result of the confluence of many technologies, including object-oriented programming, distributed computing, and artificial life. Thousands of species of animals and plants inhabit the portion of the sea where the tide is so strong that it refreshes the environment. In a similar manner, agent technology provides many interesting "foods" for researchers that result from this new interaction between different disciplines.

Why can the agent technology retain coherency instead of dispersion as a result of strong perturbations? It is because the goal of agent technology is intuitively very clear to us even though there is a technical difficulty in defining the agent-hood. The goal of agent technology is to design a new type of artifact that can cohabit with us. It is dramatically different from designing tools. It is more like characterizing dependable partners. The advance of information and communication technology has made this goal feasible.

The most important aspect of the agent technology is that it encompasses intelligence and interaction. It successfully integrates two separate disciplines, Artificial Intelligence (AI) and Human Computer Interaction (HCI), which should have been studied together; there was an unfortunate gulf between the two. AI has placed emphasis on developing artifacts with intelligence. A significant amount of AI research effort has been spent on techniques aimed at knowledge-based systems or learning machines, but little has been spent on designing interaction between humans and artifacts. Quite the opposite holds true for HCI research.

The essence of agent technology is mediated interaction between humans and artifacts. The broker agents mediate interaction among software agents, ranging from information brokerage to more complex negotiations such as commercial trade. The interface agents or embodied conversational agents mediate interaction among people by telling stories meaningful to the participants of the conversation.

Interaction enables intelligence. Intelligence emerges from interaction. Humans are intelligent not because their individual brain cells are clever. Instead, their intelligence comes from the interaction of billions of brain cells. Similar observation might hold in a macroscopic view. It is very hard for a human to become intelligent if she or he lives alone in a universe. Instead, her/his intelligence is formed from interaction with other people. Meanwhile, intelligence adds value to interaction. Interaction may not be interesting without intelligence. Intelligent members may allow their community to be intelligent as well, even though it is more a necessary condition than a sufficient condition.

The most important application area for the agent technology is the information space that is exponentially growing along with the advance of the information and communication technology. We look forward to truly dependable agent technology that can help us enjoy the adventures in this new universe.

Toyoaki Nishida
The University of Tokyo

Preface

Over the past two decades we have experienced a revolution in information availability and exchange. The amount of information available is growing at an exponential rate. With sensor technology for example, consider the European Space Agency's ERS-1 Space-based Synthetic Aperture Radar. This crosses the continent of Australia in few minutes, sending down imagery at a rate of 108Mb/s. There are other hyperspectral sensors with an even greater capacity for collecting information. In the communications area, the number of mobile telephone subscribers was about 50 million world-wide in 1994. This figure will approach 600 million this year. We read in the popular media that the number of internet sites goes up by thousands each day. Global shrinkage caused by the Internet is the single most important and evolutionary feature of this decade. We certainly live in an information rich environment. A quantum change is occurring in our society as far as the information is concerned.

Information agents are one of the most promising tools to conduct business in this information rich world. Information agents are the software systems that are capable of performing intelligent tasks in dynamic and unpredictable environment. It is desirable to incorporate in agents the characteristics such as autonomous, adaptive, collaborative, communicative, mobile, and reactive.

There are many real world problems in the world including decision making and classification which do not provide needed information, or the systems under consideration are not well defined. These problems are not easy to solve using conventional computing approaches but intelligent agent paradigms may play a major role in these areas.

The purpose of this book is to present the research results to the applications researcher rather than the agent purist. This book covers a broad selection of applications that will serve to demonstrate the applications of agents in this information rich society.

We believe that this book will be of great value to the researcher and scientists/engineers alike. The research students will receive an in-depth tutorial on the topics covered.

We are grateful to the authors for their interesting chapters. We wish to thank the reviewers for their time and expertise. We are indebted to Berend Jan van der Zwaag for his help in the preparation of the manuscript. Thanks are due to the editorial staff of the Springer-Verlag for their excellent editorial assistance.

L.C. Jain
Z. Chen
N. Ichalkaranje

Contents

Chapter 1.
Agent-based intelligent information dissemination in dynamically changing environments
H. Sevay and C. Tsatsoulis

Chapter 2.
Automating human information agents
S. Franklin

Chapter 5.
Software agents for Internet-based systems and their design
H.H. Pham

Chapter 6.
Compositional design and maintenance of broker agents
C.M. Jonker and J. Treur

Chapter 7.
Collective behavior evolution in a group of cooperating agents
J. Liu and J. Wu

Chapter 8.

Applications of information agent systems

M. Klusch, H.-J. Bürckert, P. Funk, A. Gerber, and C. Russ

Chapter 9.

The use of virtual worlds and animated personas to improve healthcare knowledge and self-care behavior: the case of the Heart-Sense game

B.G. Silverman, J. Holmes, S. Kimmel, C. Branas, D. Ivins, R. Weaver

Chapter 10.
Using agents to build a practical implementation of the INCA (Intelligent Community Alarm) system
M. Beer, W. Huang, and A. Sixsmith

Chapter 1

Agent-Based Intelligent Information Dissemination in Dynamically Changing Environments

H. Sevay and C. Tsatsoulis

This chapter presents an agent-based approach to intelligent information dissemination in dynamic environments. We will describe a prototype agent-based system, called the Anticipator, which enables an information dissemination system to adapt to changing information needs of users and provide the users with critical information relevant to those dynamic changes. The Anticipator models the information needs of each user with a dynamically adaptable profile that describes the current information needs of that user and how to adapt those information needs according to the changes in the environment. The Anticipator employs two types of agents, Profile Agents and Event Monitoring Agents. Profile Agents manage runtime user profiles. Event Monitoring Agents watch for interesting changes in the environment and report such changes to the Profile Agents such that the content and the frequency of the information sent to users can be adapted dynamically.

1 Introduction

Information dissemination via the "pull" technology defines much of today's information user's activity. Gathering information using web search engines is one very common example of this approach. However, the "pull" technology assumes that the individual user knows about what data sources are available, how frequently to search these data sources for appropriate information or potential updates as well as how to further filter the information gathered from these data sources after all the processing is done by the underlying information gathering tools. Therefore information "pull" can be considered a form of exploration through an information space, and, as such, it is not a reactive or an anticipatory mechanism for information collection. Moreover, in "pull" technology, there is a heavy burden on the user to collect the needed information. To have the most appropriate information in a dynamic setting, users need to have access to all necessary data sources to retrieve the needed information and monitor the changes in data to detect interesting events, and then link these changes to their own information needs continually. As a result of this burden on the user, not all types of information needs can be satisfactorily met via information "pull".

The capability to anticipate the information needs of users is critical for building information dissemination systems that operate in dynamic environments. As the environment changes, the information needs of users may change dynamically. Therefore, information dissemination systems need to adapt to these changes to provide the most appropriate information to their users at any given time by reacting to interesting changes in the world and, whenever possible, by anticipating future information needs based on the current state of the world.

Contrary to information "pull", in information "push" the burden is on the system for delivering the needed information to the user at the appropriate time. The user is assumed to be a direct consumer of information, and the information is assumed to be delivered automatically, without the user being involved in the information collection task. In this second approach, it is, therefore, the function of the system to know what information will be needed by a given user at what time, how frequently to deliver this information, and, most importantly, how to adapt to the dynamic changes in the environment as these changes lead to changing information needs. Therefore a mechanism for anticipating user information needs fits naturally into the "push" paradigm.

In this chapter, we present an agent-based approach to intelligent information dissemination in dynamic environments. Particularly we focus on a prototype agent-based system, called the Anticipator, which enables an information dissemination system to adapt to changing information needs of users and provide users with critical information relevant to those dynamic changes. In the Anticipator, the information needs of each user are modeled using a dynamically adaptable *profile* that describes the current information needs of that user and how those information needs are to be adapted according to the changes in the environment.

The Anticipator employs two types of agents, a set of Profile Agents (PAs) and a set of Event Monitoring Agents (EMAs) to adapt both the content and frequency of the information sent to the users. An Event Monitoring Agent watches for an event that a user is interested in, and it reports the occurrence of this event to a particular Profile Agent. A Profile Agent maintains a parameterized model of the changing information needs of a specific user, and it adapts this model (i.e., profile) based on the dynamic changes detected by individual Event Monitoring Agents.

For each new user that needs to be added to the system, the Anticipator instantiates a profile that applies to that user. The primary components of a user profile are information requests that can be active or inactive at any given time, and event rules that describe how that profile needs to be adapted when interesting changes happen in the environment. A profile is a predefined generic template for a particular type of user, and its instantiation is based on the latest known state of the world. From this point on until the user leaves the system, that profile lives as a persistent object continually adapting the information sent to the user for whom it was instantiated. Each user profile changes dynamically as a result of events of interest in the environment that may warrant specific changes or an entire *reinstantiation* of that profile.

When a profile is instantiated for a user, there are two types of Event Monitoring Agents that are created. The Anticipator creates an EMA for each information request that applies to the current state of the user and the environment. It also creates EMAs for each event rule in the profile. Both the information requests and the event rules are instantiated based on the latest known state of the world.

There are two types of Event Monitoring Agents in the Anticipator: (1) time-driven EMAs that periodically take an action such as submitting information requests on behalf of users; (2) data-driven EMAs that watch for interesting changes in the environment and subsequently notify the appropriate Profile Agents so that the user profiles being managed by those Profile Agents can be adapted to the current situation.

Our discussion in this chapter will focus on the Anticipator that operates as an intelligent component of an overall adaptive information dissemination system. We will also discuss how dynamically adaptable user profiles persistently adapt the information needs of each user as changes occur in the environment.

The remaining sections in this chapter are organized as follows: Section 2 provides an overview of the Anticipator system, and Section 3 introduces the architecture of the Anticipator. Section 4 discusses how Profile Agents operate within the Anticipator framework. Section 5 describes the profile instantiation and reinstantiation processes. In Section 6, we describe the details of the profile information request representation. Section 7 covers the operation of the Event Monitoring Agents. Section 8 presents an example that demonstrates the fundamental operations of the Anticipator. In Section 9, we discuss other research work closely related to this study. In Section 10, we briefly discuss future work. Finally, in Section 11, we present conclusions. Please note that, throughout this chapter, we have edited the format of most of the knowledge representations and written them in pseudo form to make them easier to understand.

2 Overview of the Anticipator

The problem we are addressing is delivering information to users at the *right* time and adapting the content and frequency of the information to be delivered depending on the changes in the environment, so that the user receives only pertinent information.

To accomplish this task, an information dissemination system must have access to both the state of the user and the state of the rest of the environment that pertains to that user. To be able to adapt the information sent to the user dynamically, the information dissemination system has to monitor particular aspects of the world that are of interest to each given user. Therefore, first we need a mechanism for describing the information needs of a given user under changing conditions. Second, we need a mechanism for interacting with the data sources that can provide the user with the needed information. Third, we need an event monitoring mechanism to watch for interesting changes in the environment and alert the system of those changes, so that the information needs of the user can be adapted dynamically according to the profile.

In the Anticipator, user profiles are managed and adapted by Profile Agents that stay active in the system from the time a user is entered into the system until that user leaves the system. Each Profile Agent also has a number of Event Monitoring Agents (EMAs) associated with it. The EMAs are dynamically created based on the event descriptions in a profile. Some EMAs are responsible for periodically sending information requests to data sources so that the user receives the needed information. Some EMAs are responsible for monitoring the state of the world and alerting the Profile Agents they are linked to of these changes. The state of the world is assumed to be reflected at the data sources, and, hence, the monitoring of the state of the world happens by periodically accessing the data stored at these locations.

Because of this continual feedback that Event Monitoring Agents provide to the Profile Agents, one can view the operation of the Anticipator as a closed feedback loop. The goal of the system is to *stabilize* the output of the system as the environment changes, hence reducing the *error* between the desired output and the actual output. The output of the system is the information that is being provided to the user, and the desired output is the information that should be provided to the user under the current conditions. Therefore each Profile Agent tries to adapt the information content and the frequency with which that information is sent to the user so that user receives needed information at appropriate update intervals. Alternatively, the behavior of the Anticipator can be represented as a state machine, with the important distinction that only the current state actively runs in the system at any given time, and the remaining states lay dormant implicitly in the user profile representation without any active component in memory.

3 The Anticipator in an Information Dissemination System

Figure 1 shows how the Anticipator fits as a component in the overall architecture of an information dissemination system. The User Registry

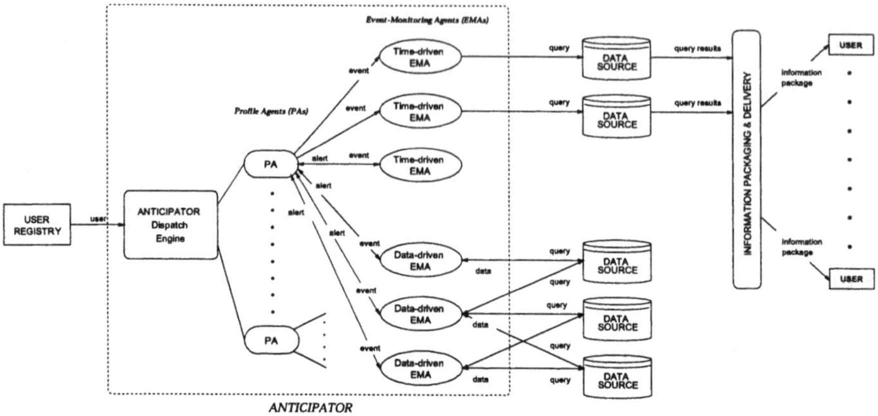

Figure 1. Anticipator Architecture as part of a larger information dissemination system.

adds and removes users from the system. The Anticipator Dispatch Engine creates Profile Agents for each new user, manages working Profile Agents, and also removes Profile Agents when users leave the system. The Data Sources provide the needed information that users require by responding to information requests that time-driven EMAs submit to them. We assume that the Data Sources continually warehouse data about the state of the world (e.g., weather conditions, medical records, locations of mobile vehicles, etc.). Data-driven EMAs watch for interesting events in the state of the world, and, when they detect such events, they immediately alert the Profile Agents they are associated with of these events. In response to these alert messages, Profile Agents dynamically modify the information content and update rate flowing to the users. As Data Sources produce data to be sent to the users, the Information Packaging and Delivery component stores the information intended for each user into an information package and sends that package to the appropriate user.

4 Profile Agents

The task of a Profile Agent is to manage the user profile of a specific user. Each user profile is initially derived from a generic profile template that is designed for a certain type of user (e.g., stock broker, military

```
Profile <profile name>

   Profile variable definitions:
      profile_variable_1  <=  instantiation_rule_1
        ...
      profile_variable_N  <=  instantiation_rule_N

   Information Requests:
      information_request_1  ...  information_request_N

   Event rules:
      event_rule_1  ...  event_rule_N
```

Figure 2. General format of a user profile.

strategist, economist, pregnant woman, computer user, etc.). Upon instantiation, a user profile becomes an active object in the system. The Profile Agent with the user profile it is managing and all of the Event Monitoring Agents associated with it live as a set of distributed elements in the Anticipator but work as a collective entity.

Figure 2 shows the general format of a user profile, which is comprised of three components:

- Profile variable definitions
- Information requests
- Event Rules

4.1 Profile Variable Definitions

Profile variables are used to parameterize different aspects of a given user profile. Their values are assigned by special rules, called instantiation rules (described next), which are referenced by each profile variable definition. There can be as many profile variables as needed in different parts of a profile.

4.2 Instantiation Rules

Each *instantiation rule* instantiates (or reinstantiates, as the case may be) the value of a specific profile variable. An instantiation rule decides on the value of a profile variable depending on the particular aspects of the

Get a satellite image
centered at location $?user\_location$
with resolution $?res$

Figure 3. An example information request for obtaining the satellite image of the area centered at user's location with a certain resolution.

current state of the world. Hence the value assigned to a profile variable may differ depending on the current conditions.

4.3 Information Requests

Information requests describe the specific information that needs to be retrieved from available data sources on behalf of users. They are parameterized to allow dynamic adaptation of the content of the information that users will receive. For example, suppose that a user needs temperature information. An information request to ask for this information can be written such that location can be a parameter so that the temperature information request can be customized to any location.

In the Anticipator, an information request can be active or inactive at any given time. Active information requests are the ones that will ask for the currently needed information for a user. Inactive information requests are the ones that do not apply to the current state of the world; but, upon changes in the environment, inactive information requests can potentially be activated. Therefore the state of the world – including that of the user – drive which information requests must be activated and which ones must be deactivated at any given time.

For example, the information request given in Figure 3 obtains a satellite image of the region centered at user's current position with a certain resolution. Let us suppose that the resolution of the image is adjusted based on the instantiation rule given in Figure 4. This example rule depends on the terrain type of the current user location, and the terrain information is retrieved from the data source that stores and updates information about the user. If, for instance, the user is in a densely forested area, the resolution of the image will be set to *high*.

```
Rule image-resolution
   if the user is currently in a desert then
      ?resolution ← low
   else if the user is currently in an open area then
      ?resolution ← medium
   else if the user is currently in a densely forested area then
      ?resolution ← high
   else
      ?resolution ← normal
```

Figure 4. Example instantiation rule that can assign different values to a profile variable based on the current state of the world.

4.4 Event Rules

An *event* is a condition that may be true of a set of data at a given point in time. Each event rule describes an event that is known to be of interest to the type of the user for which the profile was designed for, and how the current state of a user profile needs to be modified when that event condition is satisfied. Specifically, each event rule specifies a set of actions to be executed when the event condition of that rule is satisfied.

For this mechanism to work, a Profile Agent creates a data-driven Event Monitoring Agent for each event rule in a profile. The EMA then starts monitoring the condition of the corresponding event rule. When an EMA detects that its event condition is satisfied, it alerts the Profile Agent, and the Profile Agent takes the actions specified in the corresponding event rule.

Figure 5 shows the general format of an event rule, which, in structure, is very similar to the event-condition-action (ECA) rules in active databases, which also provide for any number of actions that can be executed upon the satisfaction of the event condition.

5 Profile Instantiation

In the Anticipator, profile instantiation is the very first step that takes place when a new user is being added to the system. Profile instantiation happens in two stages. For each profile template to become an active entity in the system, the profile variables in that template must first be instantiated by their instantiation rules. Next the active information re-

E: A logical statement
A_i: An action

```
if (E) then
   execute_action( A_1 );
   execute_action( A_2 );
   ...
   execute_action( A_N );
end if
```

Figure 5. General format of an event rule.

quests in the profile are instantiated now that all of the profile variables have been assigned ground values. Then, for each new user, the Anticipator creates a Profile Agent. Given a profile template for each user, the Profile Agent starts managing that profile after instantiating the profile template and creating the Event Monitoring Agents. Algorithm 1 gives the algorithm for the operation of a Profile Agent, given a profile template and a unique user ID for which that template is to be instantiated.

To instantiate the profile variables in a user profile, a Profile Agent runs the instantiation rule corresponding to that profile variable. Each instantiation rule may internally require information about the state of the world to assign a specific value to a profile variable. To get the latest state of the world, a Profile Agent asks for the values of needed (state) variables from data sources.

Once the profile variables and information requests are fully instantiated based on the latest known state of the world, then the Anticipator creates the Profile Agent that will manage this newly created working profile. This includes the initial spawning of the Event Monitoring Agents. Moreover, all further runtime modifications to this profile will be carried out by this Profile Agent.

Depending on the profile design, a set of information requests may be initially marked active. Upon the completion of the instantiation process, a Profile Agent will spawn a time-driven Event Monitoring Agent for each of these active information requests. Then, periodically, each of these EMAs will automatically send an information request to the appropriate

ALGORITHM 1 : Operate Profile Agent (P, U)

P: profile template to be used
U: unique ID of a user
V_i: a profile variable
R_i: instantiation rule associated with V_i
Q_i: an active information request in the profile
E_i: an event rule in the profile
C_i: event condition of E_i
M_i: a currently running EMA created based on E_i

// Instantiate all profile variables
for all V_i in P **do**
  $V_i \leftarrow$ Execute rule R_i;
end for

// Instantiate all active information requests and spawn time-driven EMAs
for all Q_i in P **do**
  Instantiate information request Q_i;
  Spawn EMA based on Q_i;
end for

// Spawn data-driven/time-driven EMAs for all event rules
for all E_i in P **do**
  Spawn EMA based on E_i;
end for

// Watch for event alerts and adapt the profile
while (TRUE) **do**
  for all M_i **do**
    $s \leftarrow$ Get the current status of EMA M_i;

    if (s is an *event alert*, i.e., C_i is TRUE) **then**
      Execute all actions in E_i;
    end if
  end for
end while

data sources. Then the result of these information requests will be sent by the data sources to the user.

A Profile Agent will also spawn as many data-driven EMAs as the number of event rules in the profile it is managing. Each of these EMAs will monitor the state of the world by periodically asking appropriate data sources for data that they need to check on their event condition.

We must note that it may be critical that the profile variables be reinstantiated every time a data-driven EMA alerts a Profile Agent of an event so that the user profile is modified according to the most current state of the world, and hence the user can receive the information that would be most valuable to have. In the Anticipator, the implementation of this behavior depends on the profile design. In addition, the definition of which events are interesting for a user and how the information content sent to a user needs to be adapted depends on the application, particularly, the event rules in a profile.

6 Parameterized Information Requests

An active information request in a profile is sent to a particular data source (or mediator), which will provide the needed information to the user. In the Anticipator, information requests are described with six attributes. These attributes are *frequency*, *language*, *description*, *status*, *cycle*, and *body*. The *frequency* attribute specifies how often an information request needs to be submitted to the data sources so that a user receives regular information updates; for example, every hour, every ten minutes, once a day, etc. The *language* attribute specifies the language that is used in implementing the actual request that will ask for the needed information. Therefore, it is possible to use multiple query languages within a profile (for example, SQL, Lorel, MSL, etc.). For user-friendliness, each information request also contains a *description* field to allow the profile designer to include an informal description of what information will be requested and/or any other information that would be helpful in indicating the purpose of an information request in a profile. The *status* attribute specifies whether an information request is active or inactive at a given time. Hence this attribute directly controls which information requests the Anticipator will send to data sources. The *cycle* attribute, on the other hand, specifies how many times an information request should be submitted to the data sources; for example, a cycle value of 2 says that a given information request must be submitted to the data sources twice with a period specified by the *frequency* attribute, and, upon the second submission, the information request must be deactivated. The deactivation of an information request means that the Event Monitoring Agent responsible for submitting that request to the data sources will automati-

```
Information-Request <name>
    body        ← <Formal specification of the request>
    language    ← <Language of the formal specification>
    frequency   ← <value> <unit>
    status      ← [active|inactive]
    cycle       ← [1-N|infinite]
    description ← "A textual description of the request"
```

Figure 6. General format of a profile information request.

```
Information-Request Temperature-in-city
    body        ← Get the temperature in city ?city
    frequency   ← ?mins minutes
    status      ← active
    cycle       ← infinite
```

Figure 7. Example information request that retrieves the temperature in a specified city every *?mins* minutes.

cally remove itself from the system. For persistent information requests, the *cycle* value must be set to a special value, *infinite*. Finally the *body* attribute contains the formal description of the content of the information that will be requested from data sources.

The general format of a profile information request is given in Figure 6.

Figure 7 shows an example information request (written in pseudo code) that obtains the temperature in a given city. This active information request will run indefinitely since its *cycle* attribute is set to *infinite*, and it will update the temperature every `?min` minutes. Since the location is specified using a variable (`?city`), this information request can be customized to the current location of a user during the profile instantiation/reinstantiation process.

Figure 8 shows an initially inactive information request that retrieves information about gas stations within 5 miles of the current location of a vehicle. This information request could be used in providing the driver of a vehicle with a list of closest gas stations in the case the fuel level in that vehicle falls below a certain level.

Since the information sent to the user will be updated every 5 minutes in this information request, it may be necessary that the vehicle position information (*?loc*) is also updated at least at the same rate. In the Antic-

```
Information-Request Nearest-gas-stations
  body          ← Get the list of all gas stations within
                  5 mile radius of a vehicle's current
                  location ?loc
  frequency     ← 5 minutes
  status        ← inactive
  cycle         ← 1
```

Figure 8. Example inactive information request that retrieves information about gas stations closest to a vehicle's location

ipator, this can be done by defining an event rule in the user profile that will automatically reinstantiate this information request every so often.

7 Event Monitoring Agents

Event monitoring is a critical component of the task of anticipating user information needs. The Anticipator uses two types of EMAs. Time-driven EMAs are responsible for taking periodic actions such as submitting information requests to data sources according to the frequency parameter of each information request, and data-driven EMAs are responsible for detecting whether changes in data sources satisfy the event conditions they are checking.

7.1 Event Types

In the Anticipator representation, we divide the event types into two as time-driven events and data-driven events. Similar categorizations have been used elsewhere in the literature (Decker *et al.* 1996).

7.1.1 Time-Driven Events

In the Anticipator, time-driven events depend only upon the passage of time. If the required update period has elapsed since the last occurrence of the same event, then the time-driven event is said to have occurred. In the Anticipator user profile representation, the frequency attribute of an information request specifies the period of an implicit time-driven event. For example, if the frequency value of a information request is 15 seconds, then, every 15 seconds, the event condition of the time-driven event created for updating this information request will be true.

There is also an explicitly represented time-driven event type that is akin to a periodic "ping" or a (unix) signal. These events are defined to alert Profile Agents to take specific actions regularly. For example, if we are interested in the position of a moving vehicle to decide on what kind of information to send to the driver of that vehicle, then we will have to make sure that the Profile Agent associated with that moving vehicle gets periodic updates on the position of that vehicle. One method of implementing this functionality within the Anticipator architecture is to explicitly reinstantiate information requests that depend on the vehicle movement by having a time-driven EMA alert the appropriate Profile Agent to execute reinstantiation actions on the working profile of the vehicle driver.

7.1.2 Data-Driven Events

Data-driven events, on the other hand, are dependent not on the passage of time but on the satisfaction of a logical condition involving state data. Therefore, data-driven events can be conditioned on the current values of state variables that reflect the current state of the world, trends or averages involving those variables besides other types of measurements.

We can further divide data-driven events into two subcategories as *instantaneous* and *continuous*. Instantaneous data-driven events are those with an event condition that depends on the latest value of a set of state variables. Continuous data-driven events are then those that require storage of previous value(s) in some form. For example, an event condition that checks on a running average or a trend would be classified as continuous, since prior state information would have to be stored.

The categorization of events into time-driven (periodic) and data-driven (non-periodic) have been used exclusively in active database research (Hanson and Noronha 1999). The categorization of events into time-driven and data-driven is a useful one, especially because the handling of each one requires a totally distinct implementation, with the time-driven events being simpler to implement, since they only depend on the passage of time. Data-driven events, on the other hand, can be much more complex, since they need to ask data sources for state data to be able to check the truth value of their event condition.

7.2 Time-Driven Event Monitoring Agents

The description of a time-driven EMA can be based on an information request or event rule. Since the Anticipator creates a time-driven EMA for each active information request based on the frequency parameter in that request, the Anticipator does not have an explicit event representation for EMAs that are based on information requests. The only action that such EMAs take is to submit their information requests to data sources. However, remaining types of time-driven EMAs require explicit event descriptions to allow periodic execution of actions other than submitting information requests to data sources. For example, this type of time-driven EMAs can be used for pinging the appropriate Profile Agents for periodic reinstantiation of information requests.

ALGORITHM 2 : Time-Driven Event Monitor Operation

T: information request update period (frequency)
Q: instantiated information request to be sent
D: data source that can produce results for Q
U: unique ID of a user

while (TRUE) **do**
  Send Q to D and have the results forwarded to U;
  Wait T;
end while

A time-driven EMA has a simpler working structure than a data-driven EMA. When a time-driven EMA is spawned by a Profile Agent, the frequency of that EMA attribute has a specific numerical value that may have been assigned during profile instantiation or specified as a constant at the time of the corresponding profile template design. At regular intervals specified by this frequency value, a time-driven EMA sends its already instantiated information request to the appropriate data source. How we choose which data source to send an information request to is largely a question of mediation, and this issue of mediation is beyond the scope of this chapter. In our model, we assume that all information requests are sent to addresses designated *a priori*, and that mediators decide to which specific data sources to forward information requests. Moreover, since each request contains a description of a user's destination address, the data source responding to that request can send its results directly to that user.

7.3 Data-Driven Event Monitoring Agents

Each data-driven EMA in the Anticipator is created based on an event rule in a given profile. The condition of that event rule becomes the *event* that the EMA checks for. An event is usually made up of conditions that require that certain aspects of the current state of the world be known. Therefore, an EMA asks the appropriate data sources to retrieve the values of variables, called *state variables* that are used in its event condition. This activity is performed periodically. After each update of state variables, the EMA checks whether its event condition is satisfied. If so, then it means that an event has occurred. Then the EMA automatically alerts the Profile Agent that spawned this EMA. In response to an event alert, the Profile Agent takes the actions prescribed in the action section of the event rule in the profile. If, on the other hand, the event condition is not satisfied, the cycle of information updating and condition checking continues until the EMA is removed from the system. Algorithm 3 summarizes how a data-driven EMA operates.

ALGORITHM 3 : Data-Driven Event Monitor Operation

T: state variable update period
C: event condition to check for
A: Profile Agent that spawned this event monitor
S_i: a state variable

while (TRUE) **do**
  for all S_i used in C **do**
    $S_i \leftarrow$ Retrieve the value of S_i from a data source;
  end for

  if (C is TRUE) **then**
    Send an *event alert* to A that C is now TRUE;
  end if

  Wait T;
end while

The actions that the Profile Agent takes in response to an event alert are used to adapt the profile to the changes that were deemed of interest to the user. For example, these actions can direct the Profile Agent to activate currently inactive information requests, deactivate currently active information requests, or reinstantiate a set of information requests. It is also possible that some events may require that the entire profile be reinstantiated. A reinstantiation causes all of the existing EMAs to be removed from the system. Also the profile variables in the profile are reinstantiated. After the removal of existing EMAs, the Profile Agent once again creates a time-driven EMA for each active information request, and a data-driven EMA for each event rule. Algorithm 4 gives the algorithm for this profile reinstantiation process.

ALGORITHM 4 : Reinstantiate Profile (P, U)

D_i: a data-driven EMA
T_i: a time-driven EMA
U : unique ID of a user
P : profile template to be used

$S_d \leftarrow$ Set of currently active data-driven EMAs
$S_t \leftarrow$ Set of currently active time-driven EMAs

for all D_i in S_d **do**
  Remove EMA D_i;
end for

for all T_i in S_t **do**
  Remove EMA T_i;
end for

Call *Algorithm 1*(P, U); // See page 11

8 An Example

In this section, we will present an example scenario to demonstrate how the Anticipator handles dynamic changes in an environment. This example is in the domain of battlefield awareness. We will start with the profile of a military unit in active duty. The profile is given in Figure 9.

Suppose that the military unit has a unique ID of *acomp001* and that this unit has already been added as a new user to the system. In the profile

```
Profile military-unit
   Profile variable definitions:
      ?lat  ⇐ Rule_A
      ?long ⇐ Rule_B
      ?res  ⇐ Rule_C

   Information Requests:
      Information-Request friendly-info
         body        ← Get a list of friendly units within
                        25 miles radius of user's current
                        location (?lat, ?long)
         frequency   ← 10 minutes
         status      ← active
         cycle       ← infinite

      Information-Request satellite-image
         body        ← Get a satellite image of the area
                        within 25 miles radius of user's current
                        location (?lat, ?long) with resolution ?res
         frequency   ← 15 minutes
         status      ← inactive
         cycle       ← infinite

      Information-Request hostile-info
         body        ← Get a list of hostile units within
                        25 miles radius of user's current
                        location (?lat, ?long)
         frequency   ← 5 minutes
         status      ← inactive
         cycle       ← infinite

   Event rules:
      Event-Rule enemy-close-by
         if there any enemy units within 25 miles then
            Reinstantiate this profile;
            Activate information request satellite-image;
            Activate information request hostile-info;
         endif
```

Figure 9. Example profile for a military unit.

template for this user, there are three information requests and a single event rule. The first information request is the only request that is initially activated, since its status field is marked *active.* Since the remaining two requests are inactive, the Anticipator will not create any time-driven EMAs for them at the time of the first instantiation of this profile. There is only one event rule in this profile (`enemy-close-by`) that can potentially activate both of these initially inactive information requests.

Figure 10 shows an example implementation of one of the instantiation rules used in the military unit profile. Instantiation rule `Rule_A` first re-

```
Instantiation-Rule Rule_A
   unitID ← Get unit's unique ID;
   lat    ← Get current latitude of unit unitID;
   status ← Get current status of unit unitID;

   if unit unitID is at high alert then
      offset ← 20 miles;
   else if unit unitID is at medium alert then
      offset ← 30 miles;
   else if unit unitID is at low alert then
      offset ← 40 miles;
   else
      offset ← 50 miles;
   end if

   Instantiate value (lat - offset);
```

Figure 10. Example instantiation rule.

```
Information-Request friendly-info
   body      ← Get all friendly units within a radius of
               25 miles around unit's current
               location (?lat, ?long)
   frequency ← 10 minutes
   status    ← active
   cycle     ← infinite
```

Figure 11. Information request `friendly-info` after an instantiation.

trieves a few values about a unit from data sources. Then, depending on the current status of the unit, it subtracts an offset value from the unit's current latitude. This value becomes the instantiated value for profile variable `?lat1`.

After the military unit profile is fully instantiated, the profile starts running in the system in its instantiated form, and it is managed by a unique Profile Agent created on behalf of *acomp001*. For example, the only initially active the information request, `friendly-info`, might look as shown in Figure 11 after instantiation. So when the system creates a time-driven EMA for this information request, that newly created time-driven EMAs will automatically start submitting its information request to the appropriate data source every 10 minutes. This information request will then cause that data source to send information about friendly forces in the area to user *acomp001* every 10 minutes. In addition, the system will also create a data-driven EMA for the single event rule, `enemy-close-by`, to watch whether there are any enemy units approaching.

Let's imagine that a hostile unit moves within 25 miles of the friendly unit. When that happens, the event condition of the event rule `enemy-approaching` will be true. Then the system will execute the actions of that event rule, first reinstantiating the entire working profile of user *acomp001* and then activating previously inactive information requests `satellite-image` and `hostile-info`. Since the first action calls for a complete reinstantiation of the profile, the system will first remove all the existing time-driven and date-driven EMAs associated with user *acomp001*. Then it will generate new versions of those monitors based on the reinstantiated profile. With this, another cycle of Profile Agent operation will start.

9 Related Work

The problem of selecting what information is relevant to what users can be thought of as information filtering (Belkin and Croft 1992). It may also be referred to as selective dissemination of information (SDI) (Foltz and Dumais 1992) when users submit their profiles to a system from which they expect to receive information updates. In the SIFT system, for example, profiles are made up of keywords that users need to submit to an information dissemination system (Yan and Garcia-Molina 1995). SIFT differs from the Anticipator in how it represents, uses, and updates user profiles. First, a profile in the SIFT system is a single query that a user submits to the system to get periodic updates. SIFT supports both the specification of the update period and how many times an information request should be submitted; these actions are specified by the *frequency* and the *cycle* attributes in information requests in the Anticipator. Therefore, since users need to send queries to SIFT to identify what they are interested in, profiles in SIFT are composed of a set of static queries, which can only be changed by the user instead of a higher-level user profile construct as in the Anticipator. In addition, SIFT requires that users interact directly with the system, which the Anticipator aims to abstract away.

Information requests in the Anticipator are predefined in profiles in parameterized form so that they can be customized to the current state of the world. Moreover, the user need not interact with the underlying information dissemination system to receive information or possess detailed knowledge about how to retrieve information from data sources.

To filter network news, a SIFT user can modify his/her profile by adjusting the values of the precision and recall parameters. The Anticipator, on the other hand, modifies its profiles automatically at runtime as a result of events reported to Profile Agents by data-driven Event Monitoring Agents.

Systems such as SIFT can work well depending on the quality of the retrieval algorithm they employ; however, they do not address the problem of adapting to changing user information needs. On the other hand, active database systems (Paton and Diaz 1999) provide event alerters (triggers) that notify applications of interesting data events (Dittrich *et al.* 1986, Hanson and Noronha 1999). However, an alert mechanism can only let a user know about relevant changes, but it does not fully support a user whose information needs change due to changes in an environment. Especially when we take into account rapidly changing decision environments such as battlefields, critical patient monitoring, or the stock market, it becomes apparent that we need a heavily user-centric approach for dynamic and intelligent information dissemination. We therefore believe that employing user profiles in the proactive manner that we described in this chapter will bring both increased expressibility and adaptability to information dissemination systems.

There is a clear difference between information systems that filter electronic mail, Usenet news, and the World Wide Web documents, and the kinds of information dissemination systems we focused on in this chapter. In information filtering and web browsing, the sources of information are documents. Existing information filtering systems (Papakonstantinou and Vassalos 1998, Armstrong *et al.* 1995, Sorensen and McElligott 1995, Lieberman 1995, Balabanovic 1997) represent documents using different models and they try to match user preferences saved in profiles to these models. Conversely, in the Anticipator, the information requests in profiles directly identify the information that needs to be sent to the users. That is, the Anticipator does not reason about the contents of the material (text documents, maps, images, etc.) that its requests from data sources on behalf of its users.

Some of the information filtering systems use relevance feedback and information about the actions of users (browsing, deleting files or emails, etc.) to learn and adapt user profiles. For example, the Letizia system (Lieberman 1995) only uses the actions of the users for learning, since the goal of that system is to minimize its interaction with the users. Other systems such as NewT (Sheth 1994) and Amalthaea (Moukas 1998) use genetic algorithms in conjunction with relevance feedback from users to evolve their user profiles such that those profiles are adapted according to changing interests of the users over time. Relevance feedback is also used in (Edwards *et al.* 1997) to learn user profiles based on rule induction and instance-based algorithms. Genetic algorithms are used in (Sorensen and McElligott 1995) to learn user profiles, but the approach is different from that of NewT and Amalthaea.

In recent years, search engines and web spiders have become very popular tools for locating relevant documents on the Internet, but there are numerous domains where the retrieval of very specific and potentially changing information is required. Paton and Diaz classify such systems as open database applications (Paton and Diaz 1999). One important characteristic of these systems is that they operate in dynamically changing domains such as battlefields, patient monitoring, and air traffic control. However, these systems also do not seem to be organized from the perspective of individual users with potentially changing information needs. Thus the Anticipator aims to provide a methodology for an alternatively user-centric approach.

10 Future Work

In its current status, the Anticipator does not support learning, and there is no mechanism for user-level feedback. Since an information dissemination system such as the Anticipator needs to support a wide variety of users, existing profile templates may not always be sufficient. Therefore, it is conceivable to use learning methods to identify new information needs and conditions under which those information needs would be valid.

11 Conclusions

Information systems can be more beneficial to their users if they can dynamically anticipate the needs of their users. To this end, the profile reinstantiation process is critical in providing dynamic adaptability to changes in the environment at the system level. Since the main task of the information consumer will not necessarily be information collection, having systems that automatically provide needed information to users as conditions change can be of great value in our increasingly complex information society.

In this chapter, we demonstrated that knowledge-based modeling of the information needs of users in profiles provides a strong basis for adapting to changing user information needs. The meta-level representation of user information needs that the Anticipator provides enables an information dissemination system to access a wide variety of external systems and technologies from a single interface. The benefit of a single representation system that, for example, enables different query languages for requesting information to be used, is that the generation of new profiles and editing of existing profiles can become transparent to the details of the data sources being employed, especially if we assume that mediators are used to retrieve data from different types of sources.

In the Anticipator, the actions attached to the event conditions in profiles allow dynamic adaptation of working profiles so that the information needs of users are kept consistent with the changing environment, which may frequently cause new information needs to arise. In addition, the event conditions in profiles encode *expertise* about when to send information to particular users and also what information to send. Information systems need such expertise because individual users would be overwhelmed by having to constantly link changes that occur in their environment to new information they may need. Likely they also would not possess the resources required to detect the changes that are relevant to their information needs. An automated system with a more global view of the environment can provide the needed information intelligently allowing the users to focus on their immediate goals.

Acknowledgments

This work was supported in part by DARPA contract F30602-94-C-0192 and by DARPA contract F30602-97-C-0267, both monitored by Rome Laboratories. The views, opinions, and/or findings contained in this paper are those of the authors and should not be construed as an official Air Force position, policy, or decision, unless so designated by other documentation. Many ideas in this paper were shaped with the help of Jon Dukes- Schlossberg and Yongwon Lee (Lockheed-Martin), Marc Zev and Nancy Lehrer (ISX Corporation), Alok Nigam and Ray Emami (Global InfoTek), and Hector Garcia-Molina and Joachim Hammer (Stanford University).

References

Armstrong, R., Freitag D., Joachims T., and Mitchell, T. (1995), "WebWatcher: A learning apprentice for the World Wide Web," *Proceedings of the 1995 AAAI Spring Symposium on Information Gathering from Heterogeneous, Distributed Environments*, pp. 6–12.

Balabanovic, M. (1997), "An adaptive web page recommendation service," *Proceedings of the First International Conference on Autonomous Agents*, pp. 378–385.

Belkin, N.J., and Croft, W.B. (1992), "Information filtering and information retrieval: Two sides of the same coin?" *Communications of the ACM*, vol. 35 (12), pp. 29–38.

Decker, K., Sycara, K., and Williamson, W. (1996), "Intelligent adaptive information agents," *Working Notes of the AAAI-96 Workshop on Intelligent Adaptive Agents, (Technical Report WS-96-04)*, pp. 117–126.

Dittrich, K.R., and Kotz, A.M., and Mülle, J.A. (1986), "An event/trigger mechanism to enforce complex consistency constraints in design databases," *SIGMOD Record*, vol. 15 (3), pp. 22–36.

Edwards, P., Green, C.L., Lockier, P.C., Lukins, T.C. (1997), "Exploiting learning technologies for world wide web agents," *IEE Colloquium on Intelligent World Wide Web Agents, Digest No: 97/118*, pp. 3/1–3/7.

Foltz, P.W., and Dumais, S.T. (1992), "Personalized information delivery: An analysis of information filtering methods," *Communications of the ACM*, vol. 35 (12), pp. 51–60.

Hanson, E.N., and Noronha, L.X. (1999), "Timer-driven database triggers and alerters: Semantics and a challenge," *SIGMOD Record*, vol. 28 (4), pp. 11–16.

Lieberman, H. (1995), "Letizia: An agent that assists web browsing," *Proceedings of the Fourteenth International Joint Conference on Artificial Intelligence*, pp. 924–929.

Moukas, A., and Maes, P. (1998), "Amalthaea: An evolving multi-agent information filtering and discovery system for the WWW," *Autonomous Agents and Multi-Agent Systems*, vol. 1 (1), pp. 59–88.

Papakonstantinou, Y., Vassalos, V. (1998), "Query rewriting using semistructured views," *Technical Report*, Database Group, Computer Science Department, Stanford University.

Paton, N.W., and Diaz, O. (1999), "Active database systems," *ACM Computing Surveys*, vol. 31 (1), pp. 63–103.

Sheth, B.D. (1994), "A learning approach to personalized information filtering," *Master's thesis*, MIT Media Lab.

Sorensen, H., and McElligott, M. (1995), "An online news agent," *Intelligent Agents Workshop*, British Computer Society Specialist Group on Expert Systems and Representation and Reasoning Special Interest Group, Oxford, UK.

Yan, T., and Garcia-Molina, H. (1995), "SIFT–A tool for wide-area information dissemination," *Proceedings of the USENIX Winter 1995 Technical Conference*, pp. 177–186.

Chapter 2

Automating Human Information Agents

S. Franklin

We describe a software agent technology capable of automating the entire functionality of such human information agents as insurance agents, travel agents and loan officers of banks, and many, many others. This includes negotiating in natural language, accessing databases, adhering to numerous policies, and producing products. The technology is based on a psychological theory of consciousness implemented with modules for perception, associative memory, action selection, deliberation, etc. The case study herein describes an agent whose task is to assign new billets to sailors at the end of their current duty assignment.

1 Human Information Agents

Human information agents include insurance agents, travel agents, voter registrars, mail-order service clerks, telephone information operators, employment agents, AAA route planners, customer service agents, bank loan officers, and many, many others. Such human agents must typically possess a common set of skills. These would often include most of the following:

- Communicating with clients in their natural language;
- Reading from and writing to databases of various sorts (insurance rates, airline schedules, voter roles, company catalogs, etc.);
- Knowing, understanding and adhering to company or agency policies;
- Planning and decision making (coverage to suggest, routes and carriers to offer, loan to authorize, etc);
- Negotiating with clients about the issues involved;
- Generating a tangible product (insurance policy, airline tickets, customer order, etc.).

I suspect that millions of people, mostly in developed countries, earn their livings as such human information agents. Each of these, in addition to salary and benefits, is typically provided with office space and furniture, and with computing facilities. The total yearly cost of each such agent in a developed country must conservatively be on the order of $100,000. Thus the total yearly cost of human information agents around the world must push a trillion dollars.

And, what about the cost of producing such a human information agent? A child must be produced and raised. Add to this the cost of twelve to fifteen years of formal education. Then there's on the job training. Each of these information agent jobs is relatively knowledge intensive. There's much to be learned, and that costs.

It would seem that human information agent jobs offer fertile ground for automating. How about computer programs (intelligent software agents) that would completely take over these human information agent jobs including communicating in natural language, consulting databases, adhering to policies, planning and decision making, negotiating with clients, and producing tangible products? How about taking the human out of the loop completely? The economic benefit would be enormous. From the societal side, people would be freed for more interesting, more challenging occupations.

But, you object, would people be willing to deal with a software agent instead of another person? Some would, some wouldn't. As time passed more and more people would. I now regularly use an automated bank teller, though at first I resisted. The same is true for checking on the arrival time of a flight by phone with no person on the other end. I even order a few items over the web (books and coffee). As in these instances, with familiarity would come acceptance.

OK, you say, but there's still the problem of creating such software information agents. How are you going to do that? Natural language, constraint satisfaction, planning and decision making, and negotiation aren't easy for computers. And each of these jobs is a knowledge intensive task. Can we build such software information agents?

I hope so. With funding from the United States Navy, the "Conscious" Software Research Group at the University of Memphis is in the process of designing and implementing one such software information agent. This chapter will be devoted to a description of the technology being developed. If successful, this same technology should permit the development of software agents that can perform all the tasks of several kinds of human information agents.

That's not to say that with the technology in hand the task will be easy. Each such software information agent will require a considerable knowledge engineering effort and, likely, a training and development period as a human would. Learning methods to make such development possible will be built into the agents.

The technology to be described is based on a psychological theory of consciousness and cognition. The idea is, if you want smart software agents, copy them after humans.

2 Autonomous Agents

Artificial intelligence pursues the twin goals of understanding human intelligence and of producing intelligent software and/or artifacts. Designing, implementing and experimenting with autonomous agents furthers both these goals in a synergistic way. In particular, designing and implementing within the constraints of a theory of cognition can further the first goal by providing conceptual and computational models of that theory. An *autonomous agent* (Franklin & Graesser 1997) is a system situated in, and part of, an environment, which senses that environment, and acts on it, over time, in pursuit of its own agenda. In biological agents, this agenda arises from evolved in drives and their associated goals; in artificial agents from drives and goals built in by its creator. Such drives which act as motive generators (Sloman 1987), must be present, whether explicitly represented, or expressed causally. The agent also acts in such a way as to possibly influence what it senses at a later time. In other words, it is structurally coupled to its environment (Maturana 1975, Maturana et al. 1980).

Biological examples of autonomous agents include humans and most animals. (See Figure 1.) Non-biological examples include some mobile robots, and various computational agents, including artificial life agents, software agents and many computer viruses. We'll be concerned with autonomous software agents, designed for specific tasks, and 'living' in real world computing systems such as operating systems, databases, or networks.

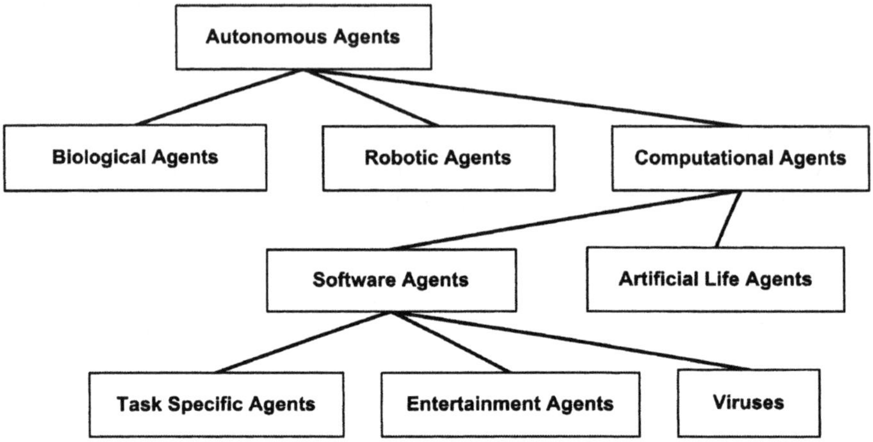

Figure 1. A taxonomy for autonomous agents.

Such autonomous software agents, when equipped with cognitive (interpreted broadly) features chosen from among multiple senses, perception, short and long term memory, attention, planning, reasoning, problem solving, learning, emotions, moods, attitudes, multiple drives, etc., are called *cognitive agents* (Franklin 1997a). 'Though ill defined, cognitive agents can play a synergistic role in the study of human cognition, including consciousness.

3 Global Workspace Theory

The material in this section is from Baars' two books (1988, 1997) and superficially describes his global workspace theory of consciousness.

In his global workspace theory, Baars, along with many others (e.g. (Minsky 1985, Ornstein 1986, Edelman 1987)) , postulates that human cognition is implemented by a multitude of relatively small, special purpose processes, almost always unconscious. (It's a multiagent system.) Communication between them is rare and over a narrow bandwidth. Coalitions of such processes find their way into a global workspace (and into consciousness). This limited capacity workspace serves to broadcast the message of the coalition to all the unconscious processors, in order to recruit other processors to join in handling the current novel situation, or in solving the current problem. Thus consciousness in this theory allows us to deal with novel or problematic situations that can't be dealt with efficiently, or at all, by habituated unconscious processes. In particular, it provides access to appropriately useful resources, thereby solving the relevance problem.

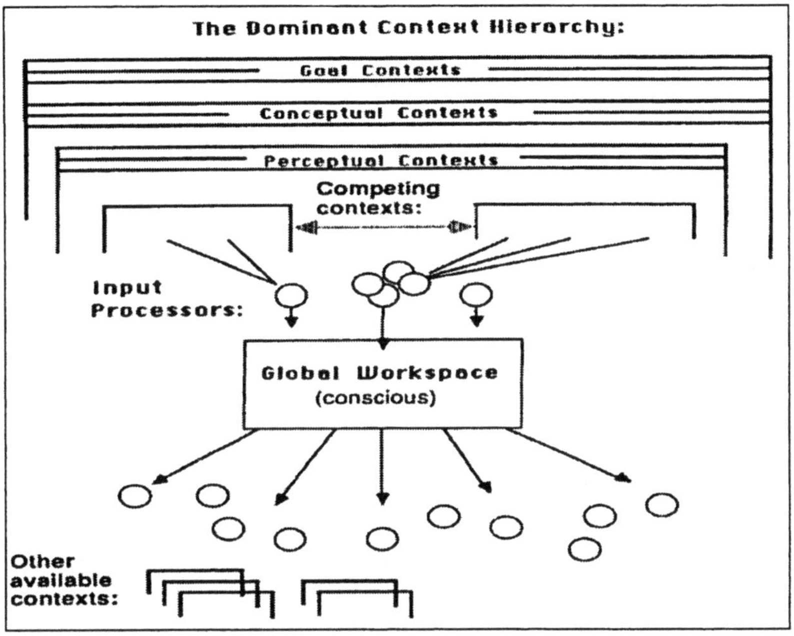

Figure 2. Global Workspace Theory.

This theory offers an explanation for consciousness being serial in nature rather than parallel as is common in the rest of the nervous system. Messages broadcast in parallel would tend to overwrite one another making understanding difficult. It similarly explains the limited capacity of consciousness as opposed to the huge capacity typical of long-term memory and other parts of the nervous system. Large messages would be overwhelming to small, special-purpose processors.

All this activity of processors takes place under the auspices of contexts (see Figure 2): goal contexts, perceptual contexts, conceptual contexts, and/or cultural contexts. Baars uses goal hierarchies, dominant goal contexts, a dominant goal hierarchy, dominant context hierarchies, and lower level context hierarchies. Each context is, itself a coalition of processes. Though contexts are typically unconscious, they strongly influence conscious processes. Baars postulates that learning results simply from conscious attention, that is, that consciousness is sufficient for learning. There's much more to the theory, including attention, action selection, emotion, voluntary action, metacognition and a sense of self. I think of it as a high level theory of cognition.

4 "Conscious" Software Agents

A "conscious" software agent is defined to be an autonomous software agent that implements global workspace theory. (No claim of sentience is being made.) I believe that conscious software agents have the potential to play a synergistic role in both cognitive theory and intelligent software. Minds can be viewed as control structures for autonomous agents (Franklin 1995). A theory of mind constrains the design of a "conscious" agent that implements that theory. While a theory is typically abstract and only broadly sketches an architecture, an implemented computational design provides a fully articulated architecture and a complete set of mechanisms. This architecture and set of mechanisms provides a richer, more concrete, and more decisive theory. Moreover, every design decision taken during an implementation furnishes a hypothesis about how human minds work. These hypotheses may motivate experiments with humans and other forms of empirical tests. Conversely, the results of such experiments motivate corresponding modifications of the architecture and mechanisms of the cognitive agent. In this way, the concepts and methodologies of cognitive science and of computer science will work synergistically to enhance our understanding of mechanisms of mind (Franklin 1997a).

5 "Conscious" Mattie

"Conscious" Mattie (CMattie) is a "conscious" clerical software agent (Franklin 1997b, McCauley, T. L. & Franklin 1998, Zhang et al. 1998b, Bogner et al. 2000). She composes and emails out weekly seminar announcements, having communicated by email with seminar organizers and announcement recipients in natural language. She maintains her mailing list, reminds organizers who are late with their information, and warns of space and time conflicts. There is no human involvement other than via these email messages. CMattie's cognitive modules include perception, learning, action selection, associative memory, "consciousness," emotion and metacognition. Her emotions influence her action selection.

Conceptually, CMattie is an autonomous software agent that 'lives' in a UNIX system. CMattie is an extension of Virtutal Mattie (VMattie) (Franklin et al. 1996, Zhang et al. 1998b, Song & Franklin 2000). VMattie, which is currently running in a beta testing stage, implements an initial set of components of global workspace theory. VMattie performs all of the functions of CMattie, as listed above, but does so "unconsciously" and without the ability to learn and to flexibly handle novel situations. CMattie adds the missing pieces of global workspace theory, including computational versions of attention, associative and episodic memories, emotions, learning and metacognition.

The computational mechanisms of CMattie incorporate several of the mechanisms of mind discussed at length in *Artificial Minds* (Franklin 1995). Each of the mechanisms mentioned required considerable modification and, often, extension in order that they be suitable for use in CMattie. The high-level action selection uses an extended form of Maes' behavior net (1989). The net is comprised of behaviors, drives and links between them. Activation spreads in one direction from the drives, and in the other from CMattie's percepts. The currently active behavior is chosen from those whose preconditions are met and whose activations are over threshold. Lower level actions are taken by codelets in the manner of the Copycat architecture (Mitchell 1993, Hofstadter & Mitchell 1994). Each codelet is a small piece of code, a little program, that does one thing. Our implementation of Baars' global workspace, discussed in more detail below, relys heavily on the playing field in Jackson's pandemonium theory (1987). All active codelets inhabit the playing field, and those in "consciousness" occupy the global workspace. Kanerva's sparse distributed memory (Kanerva 1988, Anwar et al. 1999, Anwar & Franklin submitted) provides a human-like associative memory for the agent whereas episodic memory (case-based) follows Kolodner's model (1993). CMattie's emotion mechanism uses pandemonium theory (McCauley, T. L. & Franklin 1998, McCauley, L. et al. 1999). Her metacognition module is based on a fuzzy version of Holland's classifier system (1986, 1991, 1998a).

Learning by CMattie is accomplished by a number of mechanisms. Behavior nets can learn by adjusting the weights on links as in artificial neural networks (Maes 1993). The demons in pandemonium theory become (more) associated as they occur together in the playing field of the arena (Jackson 1987). The associations that occur automatically in sparse distribute memory constitute learning (Kanerva 1988). CMattie also employs one-trial learning using case-based reasoning (Ramamurthy et al. 1998, Bogner et al. 2000).

We next turn to a brief account of how the CM-architecture uses these mechanisms to model global workspace theory. The CM-architecture can be conveniently partitioned into more abstract, high-level constructs and lower level, less abstract codelets. Higher-level constructs such as behaviors and some slipnet nodes overlie collections of codelets that actually do their work. In CMattie, Baars' "vast collection of unconscious processes" are implemented as codelets much in the manner of the Copycat architecture, or almost equivalently as Jackson's demons. Baars' limited capacity global workspace is a portion of Jackson's playing field, which holds the active codelets. Working memory consists of several distinct workspaces, one for perception, one for composing announcements, two for one-trial learning, and others.

Baars speaks of contexts as "...the great array of unconscious mental sources that shape our conscious experiences and beliefs." (1997, p. 115) He distinguishes several types, including perceptual contexts, conceptual contexts and goal contexts. The perceptual context provided by a large body of water might help me interpret a white, rectangular cloth as a sail rather than as a bed sheet. The conceptual context of a discussion of money might point me at interpreting "Let's go down by the bank?" as something other that an invitation for a walk, a picnic or a swim. Hunger might well give rise to a goal context. Contexts in global workspace theory are coalitions of codelets.

In the CM-architecture high-level constructs are often identified with their underlying collections of codelets and, thus, can be thought of as contexts. Perceptual contexts include particular nodes from a slipnet type associative memory à la Copycat (similar to a semantic net), and particular templates in workspaces. For example, a message-type node is a perceptual context. A node type perceptual context becomes active via spreading activation in the slipnet when the node reaches a threshold. Several nodes can be active at once, producing composite perceptual contexts. These mechanisms allow “conscious” experiences to trigger “unconscious” contexts that help to interpret later “conscious” events. Conceptual contexts also reside in the slipnet, as well as in associative memory. Goal contexts are implemented as instantiated behaviors in a much more dynamic version of Maes' behavior nets. They become active by having preconditions met and by exceeding a time variable threshold. Goal hierarchies are implemented as instantiated behaviors and their associated drives. (My hunger drive might give rise to the goal of eating sushi. The first behavior toward that goal might be walking to my car.) The dominant goal context is determined by the currently active instantiated behavior. The dominant goal hierarchy is one rooted at the drive associated with the currently active instantiated behavior.

Recruitment of coalitions of processors is accomplished by attention codelets, as well as by the associations among the occupants of the global workspace, via pandemonium theory. Always on the alert, attention codelets jump into action when novel or problematic situations occur. Attention is what goes into the global workspace from perception, and from internal monitoring. It also uses pandemonium theory, but requires an extension of it. Both recruitment and attention are modulated by the various context hierarchies. Learning occurs via several mechanisms: as in pandemonium theory, as in sparse distributed memory, as in behavior nets, by extensions of these, and by other mechanisms. High-level action selection is provided by the instantiated behavior net. At a low level, CMattie follows the Copycat architecture procedure of temperature controlled (here emotionally controlled), parallel terraced scanning. Problem solving is accomplished via “conscious” recruitment of coalitions of “unconscious” codelets.

Figure 3 gives a functional overview of most of the CMattie architecture. Several important functions, for example conceptual and behavioral learning, are omitted from the diagram, but not from our discussion. Detailed descriptions of the architecture and mechanisms are given in a series of papers by members of the “Conscious” Software Research group (Franklin et al. 1996, McCauley, T. L. & Franklin 1998, Ramamurthy et al. 1998, Zhang et al. 1998a, 1998b, Franklin &

Graesser 1999, Bogner et al. 2000, Song & Franklin 2000, Anwar & Franklin submitted).

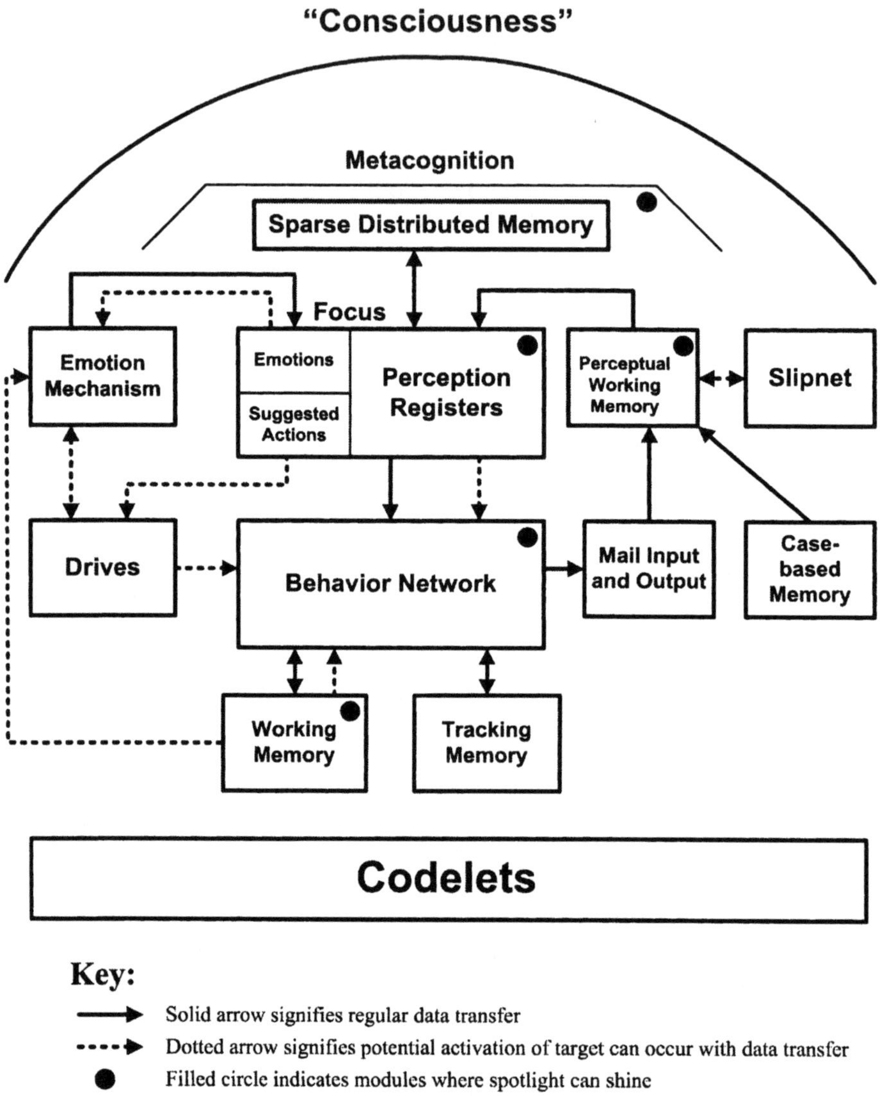

Figure 3. Most of the "conscious" Mattie architecture.

6 IDA

IDA (Intelligent Distribution Agent) is a "conscious" software information agent being developed for the US Navy (Franklin et al. 1998). At the end of each sailor's tour of duty, he or she is assigned to a new billet. This assignment process is called distribution. The Navy employs almost 300 people, called detailers, full time to effect these new assignments. IDA's task is to facilitate this process, by playing the role of detailer. She is to automate the detailer's task. Designing IDA presents both communication problems, and action selection problems involving constraint satisfaction. She must communicate with sailors via email and in natural language, understanding the content and producing life-like responses. Sometimes she will initiate conversations. She must access a number of databases, again understanding the content. She must see that the Navy's needs are satisfied, for example, the required number of sonar technicians on a destroyer with the required types of training. In doing so she must adhere to some ninety policies. She must hold down moving costs. And, she must cater to the needs and desires of the sailor as well as is possible. This includes negotiating with the sailor via an email correspondence in natural language. Finally, she must write the orders and start them on the way to the sailor.

The IDA architecture consists of both an abstract level (containing such entities as behaviors, message type nodes, metacognitive actions, etc.), and a lower, more specific level (implemented by small pieces of code). At the higher level the architecture is quite modular with module names often borrowed from psychology (see Figure 4). There are modules for Perception, Action Selection, Associative , Episodic Memory, Emotions, Metacognition, Learning, Constraint Satisfaction, Language Generation, Deliberation, and "Consciousness." As with CMattie above, many of their mechanisms were inspired by ideas from the "new AI" (the copycat architecture (Mitchell 1993, Hofstadter & Mitchell 1994, Sloman 1999), behavior nets (Maes 1989), sparse distributed memory (Kanerva 1988), pandemonium theory (Jackson 1987), and fuzzy classifier systems (Zadeh 1965, Holland 1986)). Others come from more classical AI (case-based reasoning (Kolodner 1993)).

In the lower level of the IDA architecture the processors postulated by global workspace theory are implemented, as in CMattie, by codelets, small pieces of code. These are specialized for some simple task and often play the role of demon waiting for appropriate condition under which to act. Most of these codelets subserve some high level entity such as a behavior or a slipnet node or a metacognitive action. Some codelets work on their own performing such tasks as watching for incoming email and instantiating goal structures. An important type of the latter is the attention codelets who serve to bring information to "consciousness." Codelets do almost all the work, making IDA is a multi-agent system.

6.1 Perception

IDA senses text, not imbued with meaning, but as primitive sensation as for example the pattern of activity on the rods and cones of the retina. This text may come from email messages, a chat room environment, or from a database record. Her perception module (much like that of an earlier such agent, VMattie (Zhang et al. 1998b)), employs analysis of surface features for natural language understanding (Allen 1995). It partially implements perceptual symbol system theory (Barsalou 1999), which is used as a guide. Its underlying mechanism constitutes a portion of the Copycat architecture (Hofstadter & Mitchell 1994).

The perceptual/conceptual knowledge base of IDA takes the form of a semantic net with activation passing called the slipnet (see Figure 5). The name is taken from the Copycat architecture. Nodes of the slipnet constitute the agent's perceptual symbols. Pieces of the slipnet containing nodes and links, together with codelets whose task it is to copy the piece to working memory. The codelets recognizing various concept, since they are able to simulate a concept's common forms, constitute Barsalou's perceptual symbol simulators. The slipnet embodies the perceptual contexts and some conceptual contexts from global workspace theory. There's a horde of perceptual codelets that descend on an incoming message, looking for words or phrases they recognize. When such are found, appropriate nodes in the slipnet are activated, This activation passes around the net until it settles. An idea type node (or several) is selected by its high activation, and the appropriate template(s) filled by codelets with selected items from the message. The information thus created from the incoming message is then written to the perception registers in the workspace (to be described below), making it available to the rest of the system.

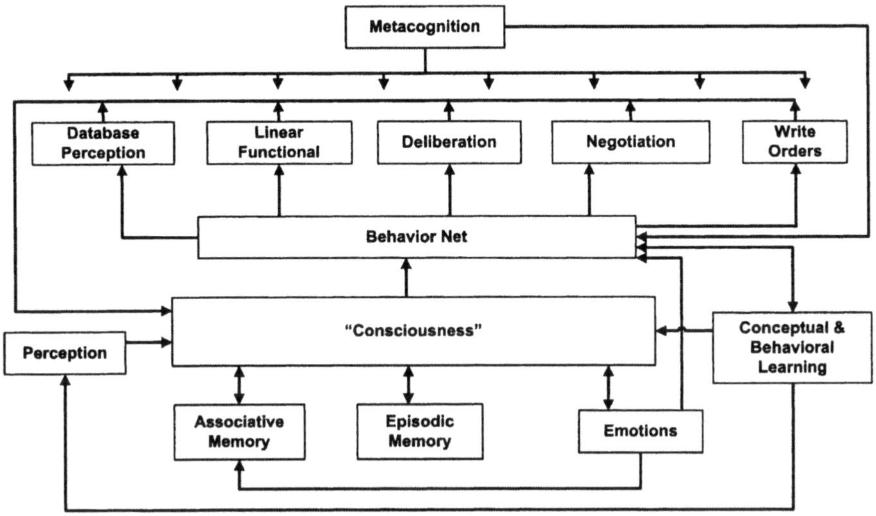

Figure 4. The IDA architecture.

The results of this process, information created by the agent for its own use (Franklin 1995), are written to the workspace (short term memory, not to be confused with Baars' global workspace). Almost all IDA's modules either write to the workspace, read from it, or both. The focus, to be described below, is part of this workspace.

6.2 Associative Memory

IDA employs sparse distributed memory (SDM) as her major associative memory (Kanerva 1988). SDM is a content addressable memory that, in many ways, is an ideal computational mechanism for use as a long-term associative memory. Being content addressable means that items in memory can be retrieved by using part of their contents as a cue, rather than having to know the item's address in memory.

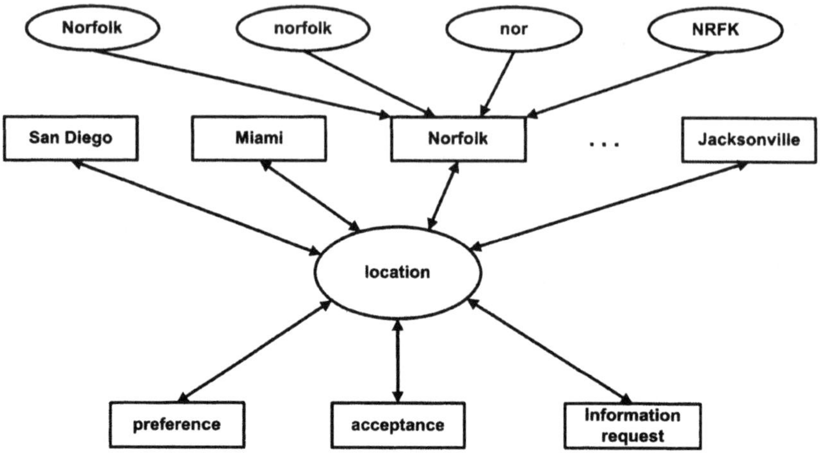

Figure 5. A small portion of IDA's Slipnet.

The inner workings of SDM rely on large binary spaces, that is, spaces of vectors containing only zeros and ones, called bits. These binary vectors, called words, serve as both the addresses and the contents of the memory. The dimension of the space determines the richness of each word. These spaces are typically far too large to implement in any conceivable computer. Approximating the space uniformly with a possible number of actually implemented, hard locations surmounts this difficulty. The number of such hard locations determines the carrying capacity of the memory. Features are represented by one or more bits. Groups of features are concatenated to form a word. When writing a word to memory, a copy of the word is placed in all close enough hard locations. When reading a word, a close enough cue would reach all close enough hard locations and get some sort of aggregate or average out of them. Reading is not always successful. Depending on the cue and the previously written information, among other factors, convergence or divergence during a reading operation may occur. If convergence occurs, the pooled word will be the closest match (with abstraction) of the input reading cue. On the other hand, when divergence occurs, there is no relation -in general- between the input cue and what is retrieved from memory.

SDM is much like human long-term memory. A human often knows what he or she does or doesn't know. If asked for a telephone number I've once known, I may search for it. When asked for one I've never known, an immediate "I don't know" response ensues. SDM makes such decisions based on the speed of initial convergence. The reading of memory in SDM is an iterative process. The cue is used as an address. The content at that address is read as a second address, and so on until convergence, that is, until subsequent contents look alike. If it doesn't quickly converge, an "I don't know" is the response. The "on the tip of my tongue phenomenon" corresponds to the cue having content just at the threshold of convergence. Yet another similarity is the power of rehearsal during which an item would be written many times and, at each of these to a thousand locations That's the "distributed" pare of sparse distributed memory. A well-rehearsed item can be retrieved with smaller cues. Another similarity is forgetting, which would tend to increase over time as a result of other similar writes to memory.

How does IDA use this associative memory? Reads and writes to and from associative memory are accomplished through a gateway within the workspace called the focus (see Figure 6). When any item is written to the workspace, another copy is written to the read registers of the focus. The contents of these read registers of the focus are then used as an address to query associative memory. The results of this query, that is, whatever IDA associates with this incoming information, are written into their own registers in the focus. This may include some emotion and some action previously taken. Thus associations with any incoming information, either from the outside world, or from some part of IDA herself, are immediately available. Writes to associative memory are made latter and will be described below.

6.3 "Consciousness"

The apparatus for producing "consciousness" consists of a coalition manager, a spotlight controller, a broadcast manager, and a collection of attention codelets who recognize novel or problematic situations (Bogner 1999, Bogner et al. 2000). Attention codelets have the task of bringing information to "consciousness." Each attention codelet keeps a watchful eye out for some particular situation to occur that might call for "conscious" intervention. Upon encountering such a situation, the appropriate attention codelet will be associated with the small number of codelets that carry the information describing the situation. This association should lead to the collection of this small number of codelets, together with the attention codelet that collected them, becoming a coalition. Codelets also have activations. The attention codelet increases its activation in order that the coalition might compete for "consciousness" if one is formed.

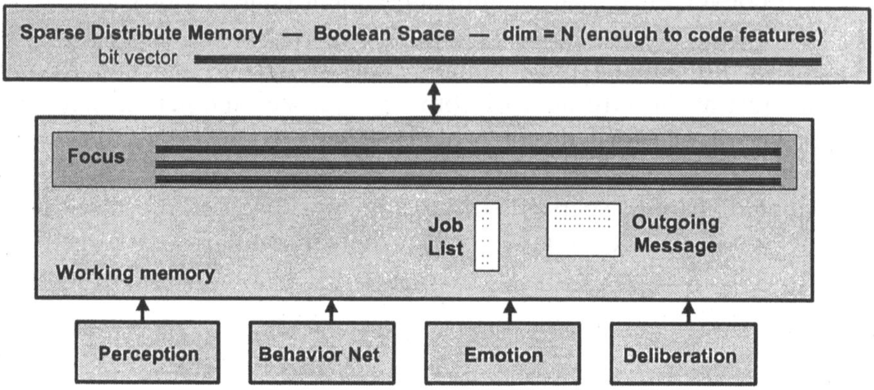

Figure 6. IDA's associative memory and workspace.

In IDA the coalition manager is responsible for forming and tracking coalitions of codelets. Such coalitions are initiated on the basis of the mutual associations between the member codelets. At any given time, one of these coalitions finds it way to "consciousness," chosen by the spotlight controller, who picks the coalition with the highest average activation among its member codelets. Global workspace theory calls for the contents of "consciousness" to be broadcast to each of the codelets. The broadcast manager accomplishes this.

6.4 Action Selection

IDA depends on expansion of the idea of a behavior net (Maes 1989) for high-level action selection in the service of built-in drives (see Figure 7). She has several distinct drives operating in parallel. These drives vary in urgency as time passes and the environment changes. Behaviors are typically mid-level actions, many depending on several codelets for their execution. A behavior net is composed of behaviors and their various links. A behavior looks very much like a production rule, having preconditions as well as additions and deletions. A behavior is distinguished from a production rule by the presence of an activation, a number intended to measure the behavior's relevance to both the current environment (external and internal) and its ability to help satisfy the various drives it serves. Each behavior occupies a node in a digraph (directed graph). The three types of links of the digraph are completely determined by the behaviors. If a behavior X will add a proposition b, which is on behavior Y's precondition list, then put a successor link from X to Y. There may be several such propositions resulting in several links between the same nodes. Next, whenever you put in a successor going one way, put a predecessor link going the other. Finally, suppose you have a proposition m on behavior Y's delete list that is also a precondition for behavior X. In such a case, draw a conflictor link from X to Y, which is to be inhibitory rather than excitatory.

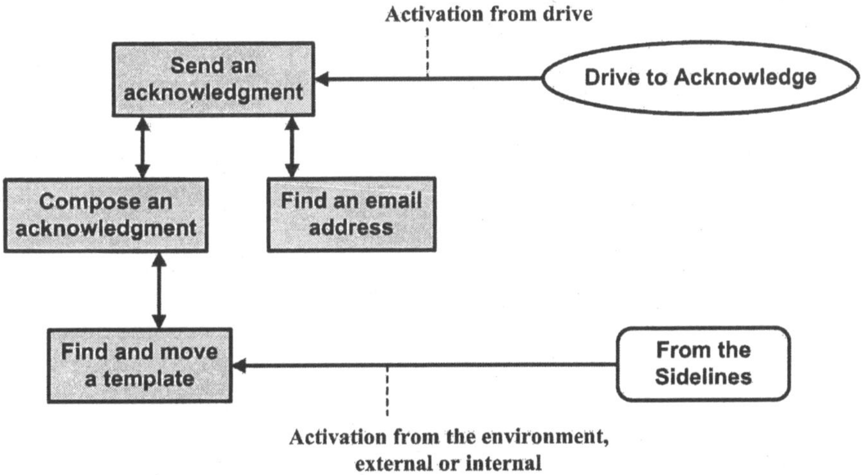

Figure 7. A simple goal structure.

As in connectionist models, this digraph spreads activation. The activation comes from activation stored in the behaviors themselves, from the external environment, from drives, and from internal states. The environment awards activation to a behavior for each of its true preconditions. The more relevant it is to the current situation, the more activation it's going to receive from the environment. This source of activation tends to make the system opportunistic. Each drive awards activation to every behavior that, by being active, will help to satisfy that drive. This source of activation tends to make the system goal directed. Certain internal states of the agent can also send activation to the behavior net. This activation, for example, might come from a coalition of codelets responding to a "conscious" broadcast. Finally, activation spreads from behavior to behavior along links. Along successor links, one behavior strengthens those behaviors whose preconditions it can help fulfill by sending them activation. Along predecessor links, one behavior strengthens any other behavior whose add list fulfills one of its own preconditions. A behavior sends inhibition along a conflictor link to any other behavior that can delete one of its true preconditions, thereby weakening it. Every conflictor link is inhibitory. Call a behavior *executable* if all of its preconditions are satisfied. To be acted upon a behavior must be executable, must have activation over threshold, and must have the highest such activation. Behavior nets produce flexible, tunable action selection for these agents.

IDA's behavior net acts in consort with her "consciousness" mechanism to select actions. Here's how it works. Suppose some piece of information is written to the workspace by perception or some other module. Vigilant attention codelet watch both it and the resulting associations. One of these attention codelets may decide that this information should be acted upon. This codelet would then attempt to take the information to "consciousness," perhaps along with any discrepancies it may find with the help of associations. The attention codelet and the needed information carrying codelets become active. If the attempt is successful, the coalition manager makes a coalition of them, the spotlight controller eventually selects that coalition, and the contents of the coalition are broadcast to all the codelets. In response to the broadcast, appropriate behavior priming codelets perform three tasks: 1) if it's not already there, an appropriate goal structure is instantiated in the behavior net. 2) wherever possible the codelets bind variables in the behaviors of that structure. 3) the codelets send activation to the currently appropriate behavior of the structure. Eventually that behavior is chosen to be acted upon. At this point, information about the current emotion and the currently executing behavior are written to the focus by the behavior codelets associated with the chosen behavior. The current contents of the write registers in the focus are then written to associative memory. The rest of the behavior codelet associated with the chosen behavior then perform their tasks. An action has been selected and carried out by means of collaboration between "consciousness" and the behavior net.

6.5 Constraint Satisfaction

At the heart of IDA's task of finding new jobs for sailors lies the issue of constraint satisfaction. Not only must IDA look out for the needs of the sailor, she must also see that the requirements for individual jobs are met, and simultaneously adhere to the policies of the Navy. Sailors tend to stay in the Navy when they are satisfied with their job assignment, and to leave at the end of an enlistment when they aren't. Thus, keeping the sailors happy is an issue of central concern to the Navy. Each individual job presents its own constraints in terms of job qualifications, location, sea or shore duty, time of arrival, etc. Finally, the policies of the Navy must be adhered to. For example, a sailor finishing shore duty should be next assigned to sea duty. Taking such issues into consideration, IDA's constraint satisfaction module is designed to provide a numerical measure of the fitness for a particular job for a particular sailor. Here's how it is to work.

Given a specified issue such as sailor preference, a particular Navy policy or specific job requirement, referred to as j for short, we define a function x_j that provides a numerical measure of the fitness of this job for this sailor with respect to this particular issue. For example, suppose the issue j is the one that says a sailor may be assigned to a job requiring a certain paygrade, if his or her paygrade is no more than one more or less. Here we might define x_j as follows: $x_j = 1$ if the sailor has the specified paygrade, $x_j = 0.5$ if the sailor's paygrade is one more or less than that specified, and $x_j = 0$ otherwise. This would provide the desired numerical measure of fitness with respect to this particular policy.

Having chosen in consultation with human detailers a collection of issues to be considered by IDA, we must create such a fitness function x_j for each of them. Computationally, the functions must be quite diverse. Most would take their input from information from the sailor's personnel record or from the job requisition list that has already been written to the workspace. As in the example above, the numerical values of the functions must lie between 0 and 1.

With these functions in hand, IDA can tell how suitable a particular job was for a specified sailor with respect to any given issue. But, what about with respect to all of them? How can we measure the overall suitability of this job for this sailor? How can we combine the individual fitness measures associated with the individual issues? What's the common currency?

What we need is, for each issue j, a numerical measure a_j of the relative importance of that issue with respect to all the other issues. Such measures can be determined in consultation with expert human detailers using statistical methods. They may also be approximated from data concerning actual assignments of sailors to jobs by human detailers. Some combination of these two methods may contribute to a more accurate choice of the a_j. Each a_j should also lie between 0 and 1, and their sum should be 1. Each a_j will be used to weight the result of the corresponding function x_j that measures the suitability of the given job for the sailor in question with respect to the issue j. Thus the weighted sum of the x_j, $\Sigma a_j x_j$, will give the required total fitness measure with respect to all the issues. This is our common currency. IDA now has a measure of the fitness of a particular billet for the sailor in question.

6.6 Deliberation in Action

IDA's relatively complex domain requires deliberation in the sense of creating possible scenarios, partial plans of actions, and choosing between them. For example, suppose IDA is considering a sailor and several possible jobs, all seemingly suitable. She must construct a scenario for each of these possible billets in order to determine whether or not a given job can meet joint temporal constraints such as the sailor's projected rotation date (PRD) and the take-up month (TUM) of the billet. And, a sailor to be assigned to a certain ship had best arrive before the ship sails. If this can't be accomplished, some other assignment must be made. In each scenario the sailor leaves his or her current position during a certain time interval, spends a specified length of time on leave, possibly reports to a training facility on a certain date, and arrives at the new billet with in a given time frame. There's travel time to be figured in. Such scenarios are valued on how well they fit these temporal constraints as well as on moving and training costs.

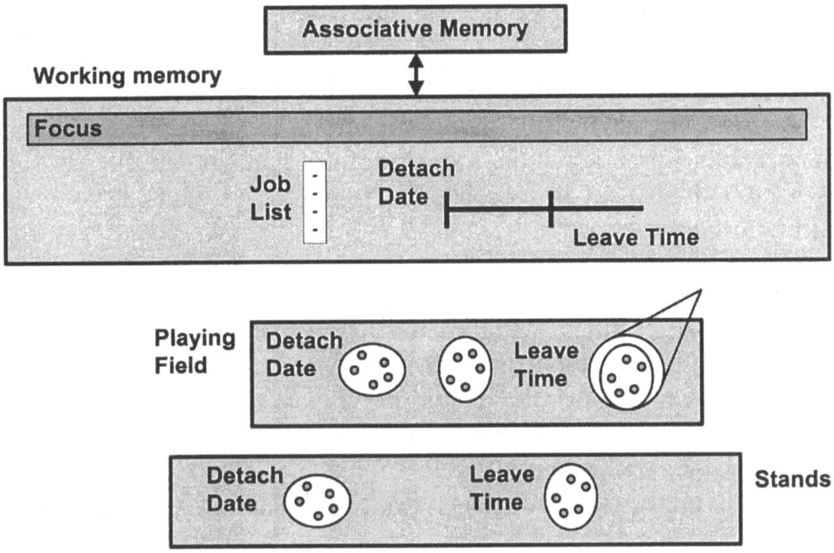

Figure 8. IDA's deliberation in action.

Scenarios are composed of scenes. IDA's scenes are organized around events. Each scene may, in theory, require objects, actors, concepts, relations, and schema represented by frames. In practice in this domain they are not all that complicated involving mostly dates and time intervals. Scenarios are constructed in the computational workspace described above, which corresponds to working memory in humans. The scenes are strung together to form scenarios. The work is done by deliberation codelets. Evaluation of scenarios is also done by codelets.

Here we'll describe the beginnings of the construction of such a scenario. The process described is common to almost all of IDA's action selection. Four type of codelets are involved, all of whom we've met before. The attention codelets serve to actuate the bringing of information from the workspace to consciousness. Information codelets actually carry most of the information during this process. Behavior priming codelets respond to "conscious" broadcasts instantiating goal structures in the behavior net (if not already there), binding variables within individual behaviors, and sending activation to relevant behaviors. Behavior codelets execute task needed to implement the behavior they serve when that behavior is selected by the behavior net to be acted upon.

On to scenario building. At this point in IDA's search for a job for the given sailor, a list of jobs coarsely selected from the current requisition list are already in the workspace. One by one they've been acted upon by the constraint satisfaction module resulting in an attached numerical fitness value. Some attention codelet notices that the last fitness value has been written next to its job. This is its cue to begin the scenario building process.

This attention codelet selects a job for the scenario (typically the one with the highest fitness) and recruits information codelets to carry specific information about the job. All these codelets are now active and, thus, available to the coalition manager. Typically they will comprise a coalition. If (or when) this coalition has sufficient activation, the spotlight will shine upon it. It's contents are then broadcast to all the other codelets. "Consciousness" has done its work.

Appropriate behavior priming codelets respond. Some extract information from the broadcast. Others know which goal structure to instantiate, in this case a create-scenario goal structure. The goal structure is instantiated, the behavior variables are bound where possible, and activation is sent to the behavior that should begin the scenario creation. Eventually, that behavior will be selected by the behavior net to be acted upon. It's codelets will then execute their tasks, writing the first scene of the scenario to the workspace. In this case the first scene will consist of a month during which the sailor is to detach from his current job.

Now the process starts again. A different attention codelet notices that the detach month of a scenario has been written in the workspace. It's its turn to recruit information to help build the next scene. The action of the "consciousness" module, the behavior priming codelets, the behavior net and the behavior codelets proceed as before. (Note that the behavior priming codelets will not have to instantiate a new goal structure this time.) All this results in a leave time period (typically thirty days) being written to the workspace as the second scene in the scenario.

The same process continues over and over again writing a scene for travel time, for proceed time (if needed and which I won't explain) for the beginning date of a training class (if needed), for the time interval of the class (if needed), and for the report-no-later-than date. The next scene written computes the gap, which depends on the relationship between the report date and the take up month. If the former is within the later, the gap is zero, otherwise more. The computation is performed by a behavior codelet.

At this point several attention codelets may vie for the next broadcast. The create-scenario attention codelet will have chosen another job for the next scenario and recruited information codelets. If the gap is non-zero, the adjust-the-gap attention codelet will try to instigate the building of a new scenario for the same job with a different detach date that may produce a smaller gap. Or, a proposer attention codelet may like this job and want to propose that it be one of those offered to the sailor (more of this in the next section).

Our contention, and that of global workspace theory, is that we humans deliberate via just such a process. From a computer science point of view, this process is grossly inefficient. How can we justify it in a software agent? In this agent we can't, since IDA's deliberation is routine in that the same actions are performed over and over, and are "well understood" by the agent. In a more challenging domain, an agent may face novel or problematic situations in which the best course of action is not at all apparent. In such a case our inefficient process may allow the agent to deliberate about the probable results of novel courses of action, and to choose a promising one without actually acting on the others. How is such a choice made? This brings us to voluntary action.

6.7 Voluntary Action

We humans most often select actions subconsciously, that is without conscious thought about which action to take. Sometimes when I speak, I'm surprised at what comes out. But we humans also make voluntary choices of action, often as a result of deliberation. Baars argues that voluntary choice is the same a conscious choice (1997, p. 131). We must carefully distinguish between being conscious of the results of and action, and consciously deciding to take that action, that is, being conscious of the decision. I am typically conscious of my speech (the results of actions) but not typically conscious of the decision to speak. However, sometimes, as in a formal meeting, I may consciously decide to speak and then do so. The decision itself becomes conscious. It's the latter case that constitutes voluntary action.

Long back, William James proposed the ideomotor theory of voluntary action (James 1890). James suggests that any idea (internal proposal) for an action that come to mind (to consciousness) is acted upon unless it provokes some opposing idea or some counter proposal. He speaks at length of the case of deciding to get out of a warm bed into an unheated room in the dead of winter. "This case seems to me to contain in miniature form the data for an entire psychology of volition." Global workspace theory adopts James' ideomotor theory as is, and provides a functional architecture for it (Baars 1997, Chapter 6). Here we provide an underlying mechanism that implements that theory of volition and its architecture in the software agent IDA.

Though voluntary action is often deliberative, it can also be reactive in the sense of Sloman (1999), who allows for the possibility or the action selection mechanism being quite complex. Suppose that, while sitting on the couch in my living room, I decide I'd like a cup of coffee and thereafter head for the kitchen to get it. The decision may well have been taken voluntarily, that is, consciously, without my having deliberated about it by considering alternatives and choosing among them. Voluntary actions may also be taken metacognitively (by Sloman's meta-management processes). For example, I might consciously decide to be more patient in the future with my young son. That would be a voluntary metacognitive decision. The IDA model includes a metacognition module that's not discussed in this paper (Zhang et al. 1998a).

But, what about action selection decisions in IDA? Are they voluntary or not? Both kinds occur. When IDA reads as sailor's projected rotation date from the personnel database, she formulates and transmits a query to the database and accepts its response. The decision to make the query, as well as its formulation and transmission, is done unconsciously. The results of the query, the date itself, does come to "consciousness." This situation is analogous to that of almost all human actions. On the other hand, IDA performs at least one voluntary action, that of choosing a job or two or occasionally three to offer a sailor. How is this done?

In the situation in which this voluntary action occurs, at least one scenario has been successfully constructed in the workspace as described in the previous section. The players in this decision making process include several proposing attention codelets and a timekeeper codelet. A proposing attention codelet's task is to propose that a certain job be offered to the sailor. This is accomplished by it bringing information about itself and about the proposed job to "consciousness" so that the timekeeper codelet can know of it. This proposing attention codelet (and it brethren) choose a job to propose on the basis of its particular pattern of preferences. The preferences include several different issues with differing weights assigned to each. The issues typically include priority (stated on the job requisition list), gap, cost of the move, fitness value, and others.

For example, our proposing attention codelet may place great weight on low moving cost, some weight on fitness value, and little weight on the others. This codelet may propose the second job on the scenario list because of its low cost and high fitness, in spite of low priority and a sizable gap. What happens then? There are several possibilities. If no other proposing attention codelet objects (by bringing itself to "consciousness" with an objecting message) and no other such codelet proposes a different job within a span of time kept by the timekeeper codelet, the timekeeper codelet will mark the proposed job as being one to be offered. If an objection or a new proposal is made in a timely fashion, it will not do so.

Two proposing attention codelets may well alternatively propose the same two jobs several times. What keeps IDA from oscillating between them forever? There are three possibilities. The second time a codelet proposes the same job it carries less activation and so has less chance of being selected for the spotlight of "consciousness." Also, the timekeeper loses patience as the process continues, thereby diminishing the time span required for a decision. Finally, the metacognitive module watches the whole process and intervenes if things get too bad.

A job proposal may also alternate with an objection, rather than with another proposal, with the same kinds of consequences. These occurrences may also be interspersed with the creation of new scenarios. If a job is proposed but objected to, and no other is proposed, the scenario building may be expected to continue yielding the possibility of finding a job that can be agreed upon.

We hypothesize that this procedure mimics the way humans make such decisions. It provides a mechanism for voluntary action.

6.8 Language Generation

CMattie's behaviors consist almost entirely of sending email messages to seminar organizers, attendees and the system administrator. In every case these messages are composed by codelets filling out templates, that is scripts with blanks allowing them to be specialized to suit the current situation. This is even true when CMattie is in the process of learning new concepts and/or behavior via interactions with organizers (Ramamurthy et al. 1998). If, for example, she writes, "If a colloquium is different from a seminar, how is it different?" she has filled in a template adding "seminar" and "colloquium." Of course, she has to be able to understand the reply.

IDA, on the other hand, must be able to respond to quite varied email messages from sailors concerning their next assignment. Many ideas from these messages will be about standard requests and can be answered with a script. It may be that all such can be so answered. We're currently cataloging such ideas, and are producing appropriate scripts. But, what of the rare idea that isn't found in our catalog? If it's something about which IDA has no knowledge she, like a human, will not be capable of any intelligent response except, possibly, to try to learn. If IDA knows something of the subject of the idea, this knowledge will likely be found in her slipnet. An improvised response would then be created by a language generation module working from the same principles as the deliberation module described above. Improvised linguistic structures created in the workspace might be combined with scripts to produce an appropriate response. All this would be controlled by a stream of behaviors in IDA's behavior net, and would be mediated by her "consciousness" mechanism.

In particular, IDA must compose a message offering the sailor a choice of one, two or sometimes three jobs. In this situation the jobs have already been chosen and the needed data concerning them are present in the workspace. The same back and forth to "consciousness" process is used as described in the previous two sections. An attention codelet, noting that the decisions on which jobs to offer have been made, brings to "consciousness" the need for a message offering the jobs. Information codelets, responding to the content of the "conscious" broadcast, instantiate into the behavior net a goal structure to write an offering message. The codelets associated with the first behavior to be executed from that structure will write a salutation for that message appropriate to the sailor's occupation and pay grade. Another attention codelet will bring to "consciousness" the number of jobs to be offered and the need for an introductory paragraph. The responding information codelets will send activation to the appropriate behavior resulting in its codelets writing that paragraph from a built-in script into the workspace. The same process will next bring into the workspace a paragraph describing the first job to be offered. Note that different jobs will require very different such paragraphs. The appropriate information codelets for the particular kind of job in question will be recruited by the "conscious" broadcast, eventually resulting in the appropriate paragraph being written to the workspace. Some modification may be made by codelets. The same process will continue until the message is composed. This is a much more complex process than that used by CMattie. It's designed this way to accommodate the more individual nature of each message. It's a scripted language generation with modifications.

7 Conclusions

Note that the skills needed by a human detailer, and hence by IDA, are much the same as those outlined for human information agent above. If the technology being developed for IDA can successfully automate the task of the detailer, there's every reason to believe that this same technology would serve to automate the tasks of very many other human information agents.

There would, of course, be problems to overcome. For example, interacting with each new database presents it own challenges. Since every human information agent's job is knowledge intensive, each new automation presents a substantial knowledge engineering problem for the would be developer. The nature of this problem would vary greatly from one human information agent task to another, so that there would be little carryover in general. Knowledge engineering techniques developed over the past several decades for building expert systems can be expected to serve adequately to help produce various "conscious" software information agents.

But even the process of automating a single type of human information agent, say a travel agent, presents still other problems. Travel agents vary in the particular knowledge required for their jobs. Some deal with domestic travel, some with international. Some specialize in a particular group of destinations. Some specialize in charter flights, others in dealing with consolidators, and still others in group travel. Some work internally for the employees of a particular company, and must know the travel policies of the company. Each of these differences requires different knowledge on the part of the agent. This same sort of analysis can be made of almost any type of human information agent. They almost always come in a number of varieties each requiring specialized knowledge. In many cases dealing with such variety by standard knowledge engineering techniques will prove too time consuming and prohibitively expensive.

The developer may choose instead to provide a developmental period for such an agent (Franklin 2000). During such a period the automating "conscious" software agent would "observe" a human performing the desired task and learn from this observation. Such learning would also require conversations with the human about new concepts and behaviors. Technology for such a development period is now being developed (Ramamurthy et al. 1998, Negatu & Franklin 1999).

Automating human information agent tasks promises to be possible, and even practical, but not cheap. Like so many artificial intelligence projects, such an endeavor will likely require a large capital outlay, which will be more than compensated for by the huge savings over the ongoing costs of a human information agent.

References

Allen, J.J. (1995), *Natural Language Understanding*. Redwood City CA: Benjamin/Cummings; Benjamin; Cummings.

Anwar, A. and Franklin, S. (submitted), "Sparse distributed memory for 'conscious' software agents."

Anwar, A., Dasgupta, D., and Franklin, S. (1999), "Using genetic algorithms for sparse distributed memory initialization," *International Conference Genetic and Evolutionary Computation (GECCO)*, July.

Baars, B.J. (1988), *A Cognitive Theory of Consciousness*, Cambridge: Cambridge University Press.

Baars, B.J. (1997), *In the Theater of Consciousness*, Oxford: Oxford University Press.

Barsalou, L.W. (1999), "Perceptual symbol systems," *Behavioral and Brain Sciences*, 22:577–609.

Bogner, M. (1999), "Realizing 'consciousness' in software agents," Ph.D. Dissertation, University of Memphis.

Bogner, M., Ramamurthy, U., and Franklin, S. (2000), " 'Consciousness' and conceptual learning in a socially situated agent," in Dautenhahn, K. (ed.), *Human Cognition and Social Agent Technology*, Amsterdam: John Benjamins, pp. 113-135.

Edelman, G.M. (1987), *Neural Darwinism*, New York: Basic Books.

Franklin, S. (1995), *Artificial Minds*, Cambridge MA: MIT Press.

Franklin, S. (1997a), "Autonomous agents as embodied AI," *Cybernetics and Systems*, 28:499–520.

Franklin, S. (1997b), "Global workspace agents," *Journal of Consciousness Studies*, 4:322–334.

Franklin, S. (2000), "Learning in 'conscious' software agents," *Workshop on Development and Learning*, Michigan State University; East Lansing, Michigan, USA: NSF; DARPA; April 5-7.

Franklin, S. and Graesser, A.C. (1997), "Is it an agent, or just a program?: A taxonomy for autonomous agents," *Intelligent Agents III*, Berlin: Springer Verlag.

Franklin, S. and Graesser, A. (1999), "A software agent model of consciousness," *Consciousness and Cognition*, 8:285–305.

Franklin, S., Graesser, A., Olde, B., Song, H., and Negatu, A. (1996), "Virtual Mattie – an intelligent clerical agent," *AAAI Symposium on Embodied Cognition and Action*, Cambridge MA, November.

Franklin, S., Kelemen, A., and McCauley, L. (1998), "IDA: a cognitive agent architecture," *IEEE Conf on Systems, Man and Cybernetics*, IEEE Press.

Hofstadter, D.R. and Mitchell, M. (1994), "The Copycat Project: a model of mental fluidity and analogy-making," in Holyoak, K.J. and Barnden, J.A. (eds.), *Advances in Connectionist and Neural Computation Theory, Vol. 2: Logical Connections*, Norwood N.J.: Ablex.

Holland, J.H. (1986), "A mathematical framework for studying learning in classifier systems," *Physica*, 22 D:307–317. (Also in Farmer, J.D., Lapedes, A., Packard, N.H., and Wendroff, B. (eds.), *Evolution, Games and Learning*, NorthHolland (Amsterdam).)

Jackson, J.V. (1987), "Idea for a mind," *Siggart Newsletter*, vol. 181, pp. 23–26.

James, W. (1890), *The Principles of Psychology*, Cambridge, MA: Harvard University Press.

Kanerva, P. (1988), *Sparse Distributed Memory*, Cambridge MA: The MIT Press.

Kolodner, J. (1993), *Case-Based Reasoning*, Morgan Kaufman.

Maes, P. (1989), "How to do the right thing," *Connection Science*, vol. 1, pp. 291–323.

Maes, P. (1993), "Modeling adaptive autonomous agents," *Artificial Life*, vol. 1, pp. 135–162.

Maturana, R.H. and Varela, F.J. (1980), *Autopoiesis and Cognition: the Realization of the Living*, Dordrecht, Netherlands: Reidel.

Maturana, H.R. (1975), "The organization of the living: a theory of the living organization," *International Journal of Man-Machine Studies*, vol.7, pp. 313–332.

McCauley, L., Franklin, S., and Bogner, M. (2000), "An emotion-based 'conscious' software agent architecture," in Paiva, A. (ed.), *Affective Interactions*, Lecture Notes on Artificial Intelligence, LNAI 1814, Springer.

McCauley, T.L. and Franklin, S. (1998), "An architecture for emotion," *AAAI Fall Symposium Emotional and Intelligent: the Tangled Knot of Cognition*, Menlo Park, CA: AAAI Press.

Minsky, M. (1985), *The Society of Mind*, New York: Simon and Schuster.

Mitchell, M. (1993), *Analogy-Making as Perception*, Cambridge MA: The MIT Press.

Negatu, A. and Franklin, S. (1999), "Behavioral learning for adaptive software agents," *Intelligent Systems: ISCA 5th International Conference; International Society for Computers and Their Applications – ISCA*, Denver, Colorado, June.

Ornstein, R. (1986), *Multimind*, Boston: Houghton Mifflin.

Ramamurthy, U., Franklin, S., and Negatu, A. (1998), "Learning concepts in software agents," in Pfeifer, R., Blumberg, B., Meyer, J.-A., and Wilson, S.W. (eds.), *From Animals to Animats 5: Proceedings of The Fifth International Conference on Simulation of Adaptive Behavior*, Cambridge, Mass: MIT Press.

Sloman, A. (1987), "Motives mechanisms emotions," *Cognition and Emotion*, vol. 1, pp. 217–234.

Sloman, A. (1999), "What sort of architecture is required for a human-like agent?" in Wooldridge, M. and Rao, A. (ed.), *Foundations of Rational Agency*, Portland Oregon.

Song, H. and Franklin, S. (2000), "A behavior instantiation agent architecture," *Connection Science*, pp. 1–24.

Valenzuela-Rendon, M. (1991), "The fuzzy classifier system: a classifier system for continuously varying variables," *Proceedings of the Fourth International Conference on Genetic Algorithms*, San Mateo CA: Morgan Kaufmann.

Zadeh, L.A. (1965), "Fuzzy sets," *Inf. Control*, 8: pp. 338–353.

Zhang, Z., Dasgupta, D., and Franklin, S. (1998a), "Metacognition in software agents using classifier systems," *Proceedings of the Fifteenth National Conference on Artificial Intelligence*, Madison, Wisconsin: MIT Press, pp. 83-88.

Zhang, Z., Franklin, S., Olde, B., Wan, Y., and Graesser, A. (1998b), "Natural language sensing for autonomous agents," *Proceedings of IEEE International Joint Symposia on Intelligence Systems 98*, pp. 374-371.

Chapter 3

Knowledge Robots for Knowledge Workers: Self-Learning Agents Connecting Information and Skills

J. Hasebrook, L. Erasmus, and G. Doeben-Henisch

The use of a desktop computer is restricted in many ways for the ordinary person today. People are not able to cope with the exponential growth of information and the increasing speed of information growth and business processes made possible by information and communication technologies. People have lost control over the information universe or *infoverse*. Intelligent technical support, not only for information storage and retrieval, but also for information selection, process planning, and decision support is needed. It is predicted that smart and mobile computing units embedded in a variety of appliances, such as TV sets and cars, will bring computing power and the common users of these intelligent appliances closer to each other by using natural language and social skills together with computer mediated communication. A general architecture of a knowledge robot or *knowbot* is described, based on a multi-agent platform and distributed computational intelligence. Knowbots consist of self-learning artificial brains connected to input sensors and output actuators of which speech recognition and synthesis are used to connect to networks of people. They have access to other software agents and computer programs through direct access or a multi-agent platform. A newly defined partnership between people and machines equipped with knowbots are a way to keep in control of the exploding infoverse.

1 Introduction

Reasoning and simulation mechanisms of unthinkable complexity currently, will take over the control of business process planning and information exchange in future. Fourth generation robots with the capability of performing more than 30 million instructions per seconds (MIPS) will be the heart of a company's knowledge base (Moravec, 1998). This is the vision of Hans Moravec, Principal Research Scientist at the Robotics Institute and Director of the Mobile Robot Laboratory of Carnegie Mellon University, Pittsburgh, USA.

The global economy is getting accustomed to the idea of the *new economy* where the workers' knowledge, creativity and skills are a company's most important capital and competitive advantage. If only part of Moravec's vision comes true, then the relevance of business related human expertise and experience would diminish very fast. Current developments seem to support this view point: A supplier of computer storage systems reports that especially banks are consuming more storage space within six months than has been used during the last twenty years. The increasing pace in product innovation and the decreasing product life cycles depreciate knowledge and skills in maintaining these products within one to three years.

The amount of information stored worldwide in the Internet starts to exceed the amount of information that has been stored in the past 60,000+ years of human culture: It has been estimated that in the years 1972 to 1980 more information has been collected than in the past 2000 years (Hasebrook 1995). Fifty years after the publishing of the first Gutenberg bible about two million books had been published. Today, more than 3000 books are published per day, more than one million per year. Some authors, therefore, are discussing the advent of the *knowledge age*. Other authors, such as Internet expert Clifford Stoll, argue that the Internet is not more than a gigantic information garbage heap.

Recent studies show that a person is not able to remember more than one to two percent of all the information perceived in the mass media, such as radio, TV, or newspapers. According to Cyveillance, the Internet was estimated to contain about 2 billion pages in July 2000 (see http:// cyveillance.com/newsroom/pressr/000710.asp). According to http:// searchenginewatch.com/reports/sizes.html Google is the biggest search engine with 602 million indexed pages in November 2000. Because of how Google uses link data, it can actually return listings for sites it has never visited. This gives it coverage of over 1.2

billion pages minus 10% for outdated links or roughly 50% of the World Wide Web pages. Even the best text searching and indexing techniques do not deliver more than 25 percent relevant links or search results (Hasebrook, 1998). This means that an 'optimal' search process accesses a quarter of a half of the information in the Internet – and one to two percent of this information can be remembered, which means only 0.1% to 0.3% of the information in the internet can be remembered by people. Thus, it has to be stated that people have lost control over the information gathered in technical information systems.

The exponential growth of information, information access at the speed of light that leads to increase business processes speeds, and the decreasing value of relevant human knowledge force engineers and scientist to re-focus their development efforts. Accessing and retrieving information are no longer the only main concern, but proper navigating inside the infoverse and then orientating and selecting the relevant information in an intuitive way by a person is of equal importance. As computers and robots provide people with incredible capabilities to process increasing amounts of data within decreasing periods of time, it seems clear that people can only master the self-made *information overload* if engineers and scientist manage to enhance the people's skills by developing proper man-machine-interfaces (MMIs) based on natural dialogues and trusted partnerships.

The main topics of interest in this new type of MMIs are mobile, ubiquitous and selective information access, enabled by smart software agents based on multi-agent platforms, using distributed computational intelligence. This means that a knowledge application will not reside only on one machine or a sub-part of a machine, but it is a whole complex interconnected system of different machines running software agents that communicates with each other and other software sub-systems and hardware that looks like a single application on a person's cellular phone, personal digital assistant (PDA) or other mobile computing devices (MCDs). Human cultural development is at a turning point where sustainable progresses can only be made if people are able to delegate information retrieval, process planning and decision support to technical systems. People will have to decide whether they want to become garbage collectors within an information jungle, or the masters of smart agent systems which they will not simply have the detailed overview in their functioning. People will have to trust these machines similar to how they trust their fellow man.

2 If It Works, It's Not AI

Up to now, the progresses of the so-called Artificial Intelligence (AI) companies have been disappointing. A recent study of the M.I.T. about the commercial success of AI startup companies comes to the conclusion: 'If it works, it's not AI'. This assumption has been reflected in the revenues of AI corporations during the past decades as can be seen in Figure 1.

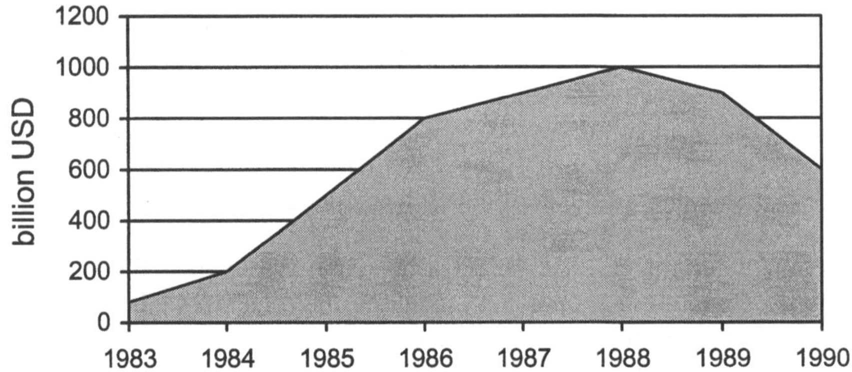

Figure 1. Approximate AI revenues 1983 – 1990 (Philipps, MIT, 1999).

The strong position of AI is to develop machines that are intelligent in a human way. On the other hand the weak position of AI is to implement programs that can be viewed as 'partly intelligent' because they are able to perform actions that used to be dedicated to human workers. This mode of AI is referred to as 'Computational Intelligence' (CI).

Patricia Churchland pointed out that strong AI position is at a stage where it tries to mimic human intelligence in the same way that the first pioneers of flight tried to mimic a bird's way of flying. No modern airplane or helicopter is flapping its wings, but relies on the principle of generating a proper lift. Using this analogy the question can be asked: What is the proper lift needed in order to lift the weak position of AI to a higher level?

A huge debate has been going on since the start of AI and continues today on what is intelligence. Out of the debate in 2000 conducted in the internet discussion group for Architectures of Intelligent Control Systems (AICS), moderated by Alexander Meystel, about a definition for what is intelligence and how can one measure the intelligence of a machine. Louwrence Erasmus proposed that intelligence are build up of intelligent factors, which all contributes to what is intuitively understood as intelligence. These intelligence factors, *IF,* are today not all known, but mathematically we can describe it as

$$Intelligence \subseteq \bigcup IF_i$$

Thus the intelligence quotient (IQ) is then defined as

$$\mathrm{IQ} \leq \int_{intelligence} m(IF)$$

where $m(IF)$ is a measure defined to measure the intelligence factor quotient (IFQ) of *IF*. Thus, in general, it can be stated that

$$\mathrm{IQ} \leq \sum \mathrm{IFQ}_i$$

In a special case where the different intelligence factors are mutually exclusive, it can be said that

$$\mathrm{IQ} = \int_{intelligence} m(IF), \text{ or } \mathrm{IQ} = \sum \mathrm{IFQ}_i$$

In the light of the definition of an intelligence factor, it can be argued that the strong position of AI have tried to build systems that, when its intelligence can be measured, will deliver the IQ of the system. Because all the intelligence factors are not known that makes up intelligence today, it is seen that there is no chance for success in determining the success of such systems. In the same light, it can be argued that the weak position of AI is trying to build systems that have a defined intelligence factor and thus also a measurable IFQ. This gives a better chance to build a system that is successful in having a specific IFQ or a combination there off.

One of the intelligent factors that is agreed upon in several scientific fields is the ability to learn. It can then be said in a general sense that if a system is intelligent it should also be able to learn; the opposite of this statement, a learning system is intelligent, is not true because of the existence of other intelligence factors that also contributes to the characteristics of intelligence. The definition of what is learning is also not generally agreed upon.

In 1998, the Institute of New Media (INM), a non-profit association, and the Bank Academy, a non-profit educational institution of the German bank associations, formed a joint venture to examine new ways of autonomous software agents which could help learners and knowledge workers in information intensive industries, such as banking and finance.

At the beginning of the year 2000, Knowbotic Systems GmbH & Co KG was founded by the inm-Network and the Bank Academy. The purpose of Knowbotic Systems is to develop and to examine knowledge robots or *knowbots* which help to fully exploit the knowledge capital of a company by facilitating information selection, planning and decision making. The mission of Knowbotic Systems relies on two basic assumptions:

- As long as key concepts, such as *learning* and *intelligence*, are not fully understood and clearly defined, computers will not be intelligent learners. A *formal learning theory* has to be deduced from recent theories and empirical studies so that a virtual test environment for knowbots can be set up which helps to measure their learning capabilities and find ways to extend them. With *learning theory* a new level of computational intelligence can be reached, that has a higher IFQ than predecessor systems.
- The critical lift for CI will not come if a system is intelligent in itself, but it will come from the capability to communicate interactively with people in a natural intelligent and social way. Thus, knowbots have to mimic intelligent communication behavior in order to transfer the results of machine intelligence to people.

An example of a natural visual interface for a knowbot is shown in Figure 2. A knowbot guides an avatar representing a human user through the city of Frankfurt. The knowbot adopts terms entered by the user and links it to its knowledge map about the buildings and other objects in the city. If an object has not been named yet, then the knowbot invents a name for the object in its own internal language. Associations between names entered by the user and the internal language gets build up by the knowbot through communicative interaction.

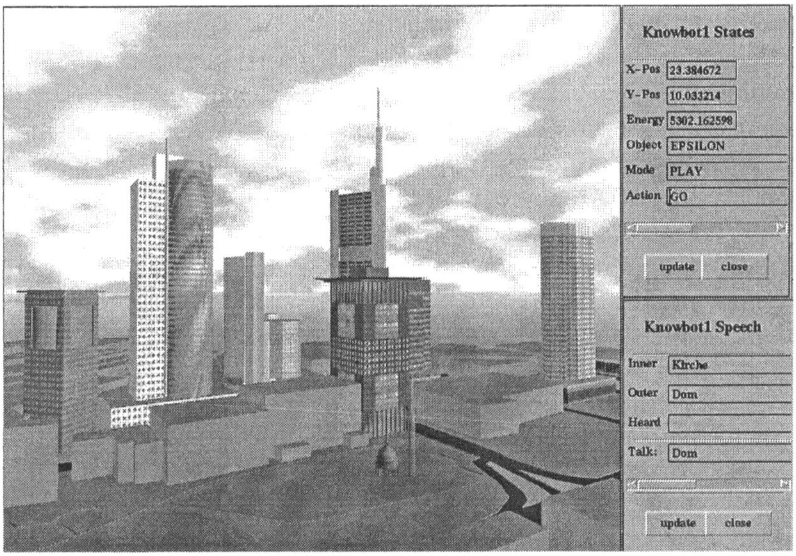

Figure 2. A knowbot guides an avatar representing a human user through the city of Frankfurt. The knowbot adopts terms entered by the user and links it to its knowledge map about the buildings and objects in the city.

3 The Artificial Brain

Most programs that mimic intelligent behavior are based on logical oriented knowledge-based techniques, which proved to be too inflexible to represent even primitive forms of learning. They elicit a number of behavioral paradoxes when used to support human learning (Hasebrook & Nathusius, 1997).

In classical AI, different forms of logical based representational schemes are used to make it easy for performing inferences (McDermott & Poggio, 1989). These representation rely on an useful representation language with the important requirement that the meaning should depend compositionally only in the meanings of the constituent structures of the sentence and not on the meaning of other sentences or surrounding context. In late 1960s, AI researchers realized that huge formal encodings of large quantities of information about a problem domain is needed to proceed to achieve sophistication in various reasoning tasks. The development of formal languages has been target to this problem of finding good representations. The main problem is that ordinary people do not communicate in formal languages that is sounding and feels foreign to the general intelligent communication act between people. These formal languages also rely on hard logic and breaks down if the naturally vague type of logic or even illogical descriptions of people are being described with these languages. Further, people do not have huge databases of formally encoded knowledge, which they carry around with them. If the brain is damaged, depending on the severity, the learning ability or other capability performance only decreases, but still such a brain can function properly with vague or imprecise results (Bloom & Larson, 1999).

In connectionism, researchers adhere to different types of artificial neural networks (ANNs) for example feed forward networks, Elman networks, time delayed networks, self organizing maps etc. (see Haykin, 1994 for more information). Figure 3 illustrates a simple feedforward neural network. ANNs have achieved some success in non-linear forecasting, pattern matching and in artificial life paradigms (see Eberhardt et.al., 1996 and Kosko, 1992). ANN exhibits the capability of redundancy and vague representation of information. It is in total contrast to classical AI. The ANNs still lack many of the vital features of biological neural networks (BNNs), such as the ability of biological neurons to allow self-modification with regard to short term and long term learning.

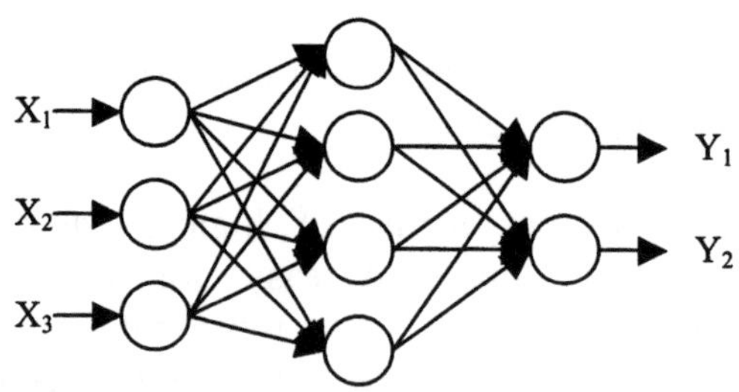

Figure 3. Example diagram of a feed forward neural network with 3 inputs, 4 hidden and 2 output neurons.

Knowbotic Systems tries to combine the behavioral perspective with the physiological perspective, both embedded in sign based communications (or semiotics) together with the concepts of learning. We call these self- learning and sign-using systems (semiotical systems) *knowbots*. Figure 4 illustrates a knowbot system.

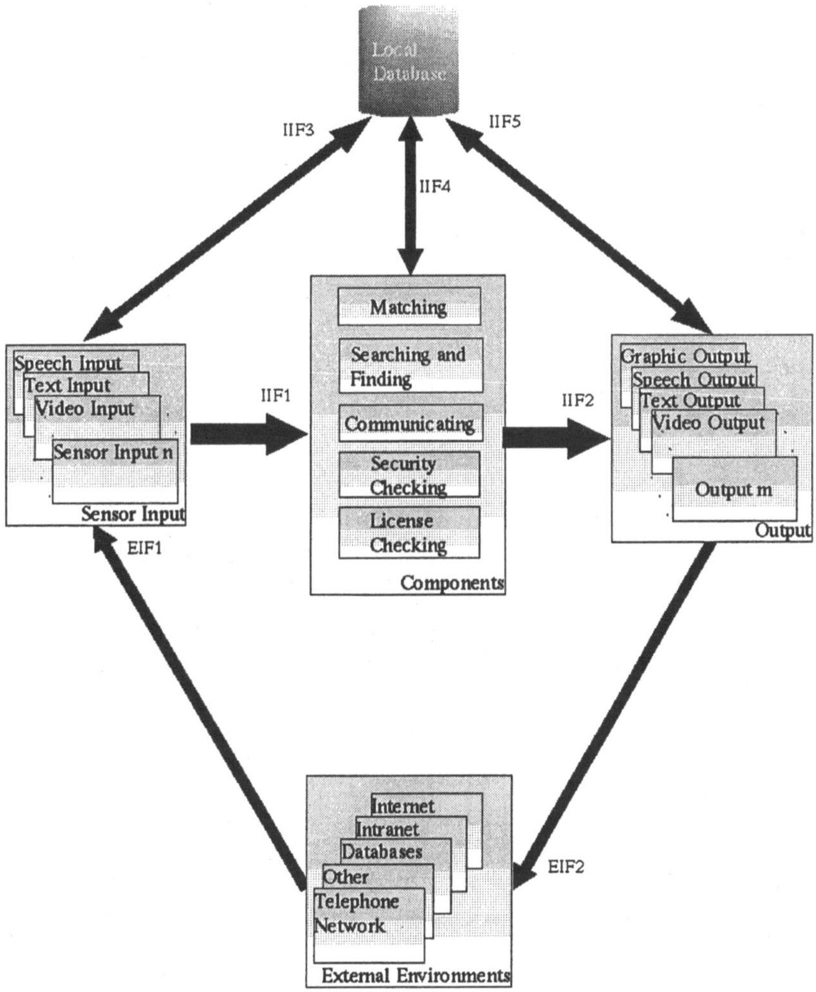

Figure 4. Diagram of a Knowbot system.

The physiological structure of the biological brain is responsible for the basic ability for observable behavior. Thus, we have to find a model of the human brain neuron, which should be empirically sounder than the classical ANNs and should be still practically feasible on standard personal computers (PCs). Knowbotic Systems' RealNeuron® simulates human brain cells with respect to the amplitude of the action potentials, the timing of the processes and the concentrations of chemical substances involved. Further, can RealNeuron® networks model local and global influence of hormones and psycho-pharmaceutics on brain cells. The difference between the RealNeuron® model and the models available in computational neuroscience is that only those properties of biological cells which are most likely underlying learning of new behavior patterns gets modeled. Figure 5 shows a block diagram of a RealNeuron® with 2 synapses on the soma, 2 synapses on the axon and two synapses on the bulb end. This example has only one output bulb end.

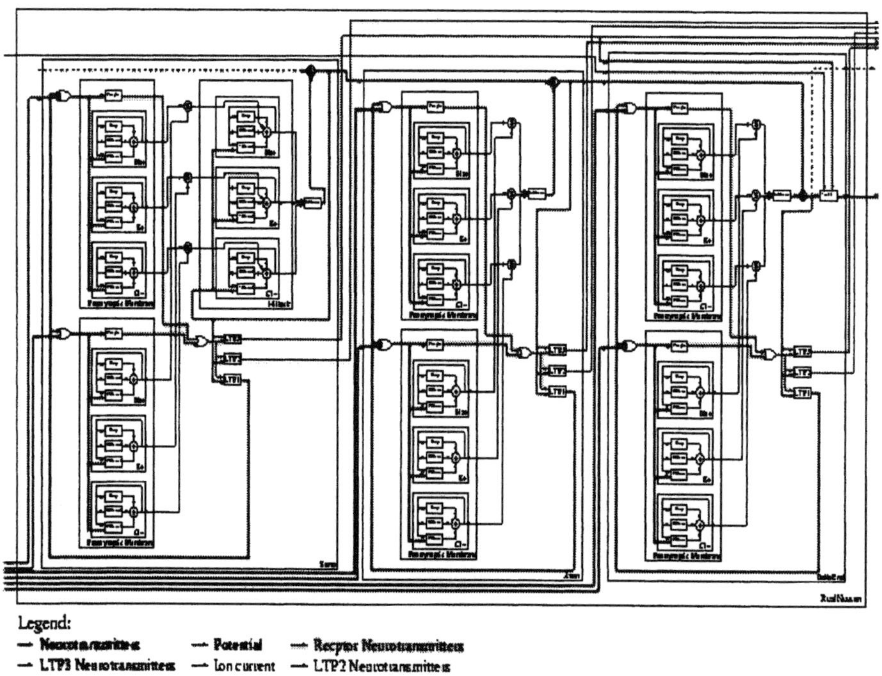

Figure 5. Example block diagram of a RealNeuron® with two synapses on the soma, 2 synapses on the axon and 2 synapses on the bulb end.

In 1950 Turing published his paper proposing the game known as the Turing test (Turing 1950, discussed in Fogel 2000). This test has proposed that you have three persons: a man (A), a woman (B), and an interrogator (C). The interrogator is separated from A and B, and sits in another room with a Teletype writer used for communications. C is allowed to ask any questions to A and B. It is the objective of C to determine which one of A and B is the man and which one is the woman. It is A's objective to deceit C to let him make the wrong decision. B's objective is to help C. Turing then replaced the original question, "Can machines think?" with the question "We now ask the question, 'What will happens when a machine takes the part of A in this game?' Will the interrogator decide wrongly as often when the game is played like this as he does when the game is played between a man and a woman?" (Turing 1950). When the interrogator shows no increased capability to decide between A and B while the machine is playing as opposed to a man, then the machine would have passed the test. Whether or not the machine should then be judged capable of thinking (intelligent) was left unanswered. Turing in fact dismissed the original question as being "too meaningless to deserve discussion". Moreover, he suggested that a thinking machine should be a learning machine, capable of altering its own configuration through a series of rewards and punishments. Thus it could modify its own programming and generate unexpected behavior.

The RealNeuron®, through the capability of simulating hormones and psycho-pharmaceutics, can change its own underlying structure dynamically. The question that still needs to be answered is, is this still enough to claim intelligence or thinking in Turing opinion or is this also just another intelligence factor that contributes to intelligence?

The restriction to the surface of meaning (i.e., to the overt observable behavior like in a Turing test) does not say too much about underlying mechanisms, which are responsible for the generation of behavior. Taking into account these underlying structures the fields of neurobiology and phenomenology are entered, which are the two main, but methodologically different, sources of information about inner states of an organism. If the additional assumption is made that the subjective experience within a phenomenological approach is caused by certain physiological body states, neurobiology becomes the primary source of information. Although neurobiological facts cannot be interpreted without a link to the actual behavior or subjective experience, they do represent the core of our knowledge about the mechanisms underlying the observable behavior.

The simulation of BNNs developed by neurobiologists does not seem to be promising because recent attempts have shown that exact simulations of neuron brain cells consumes a vast amount of computer resources (Goddard and Hood1997, reported 18.2h computing time on 5 connected sun sparc-2 workstations to simulate one second of activity of a single neuron).

Analyzing the recent neurobiological literature we were finally able to extract a model of the human brain cell which encompasses the main features of biological neuronal cells especially those which are considered to be the key features of learning (Sheperd 1998). The neuron model allows an arbitrary number of membranes and synapses, and for the membranes one can use three chemical substances simultaneously together with the necessary ionic channels, pumps, receptors, and transmitters. This chemical mechanism is used as basis for the computation of the corresponding electric potentials. Assuming a mean signaling frequency of about 100 Hz and a mean membrane density of 11 membranes per neuron, it is possible to simulate about 350 neurons in real-time with an Alpha chip working at 600 MHz. The model simulates neurons almost perfectly with respect to the height of the potentials, the timing of the processes and the concentrations of chemical substances involved. Moreover, our neural network can model the local and global influence of hormones and neuropharmaca on brain cells.

A feature of the RealNeuron® model is that within the neuron model the response function and the threshold functions are given as concrete functions simulating the given chemical mechanism of a brain cell, that is: the model simulates the mechanism of an axon followed by bulb ends with possibly several post-synaptic membranes and finally a post-synaptic membrane. Similarly, both the action potential and threshold function are realized as a membrane model which can be triggered by potentials and react according to it's own internal state.

We are aiming to represent only those properties of biological cells which are functionally important for the processing of information and especially for the learning of new behavior patterns. Along this line of research we are looking for neuronal circuits which represent possible neuronal correlates of learning and shed some light on behavioral learning concepts.

Maass (1997) classifies the existing neural network models according to the following three generations:

- *First Generation:* McCulloch-Pitts neurons, or perceptrons, and threshold gates are used to implement multi-layered perceptrons, Hopfield nets, and Boltzmann machines. They can compute Boolean functions as well as digital functions.
- *Second Generation:* Applying activation functions with a continuous set of possible output values to a weighted sum of the inputs (e.g. sigmoid functions, linear saturated functions, piecewise exponential functions) enabled the implementation of feed forward and recurrent neural networks and networks of radial basis function units. These networks can compute Boolean function, digital functions, and continuous functions with a compact domain. Moreover, they support learning algorithms based on gradient descent.
- *Third Generation:* Spiking neurons, such as integrate and fire neurons, form networks of spiking neurons in which timing of individual computational steps plays a key-role for the computation.

Maass (1997) proved that the third generation model has at least the same computational power as neural nets from the first two generations of a similar size. Furthermore, many functions require significantly *fewer* spiking neurons than neurons of the first or second generation. We have proven that the RealNeuron® model fits the definition of a spiking neuron.

Only a few BNNs underlying learning have been identified yet. As a first test case we have chosen a classical conditioning circuit and several candidates that might be responsible for operant conditioning (Menzel 1996). In first experiments, we implemented the network that represents the eye blink reflex of a rabbit as shown in Figure 6.

The network matches the neuropsychological data qualitatively (see Figure 7): If the unconditioned stimulus (US = air flow) is given slightly before the the conditioned stimulus (CS = sound), then the connection of the US and CS is learned in a few trials,. After several runs presenting CS without US the 'connection' between the stimuli is unlearned. It is re-established very quickly, if the CS and the US are presented simultaneously again (Menzel, 1996). This means, that not only the neuro-biological structure of brain cells can be simulated on a PC, but also basic learning behavior which perfectly matches empirical data.

Up to now, no clear candidate acting as an neurological correlate of operant conditioning has been identified. In order to overcome this shortcoming, we set up an (artificial) BNN with our neurons testing in defined experimental environment instances of operant conditioning.

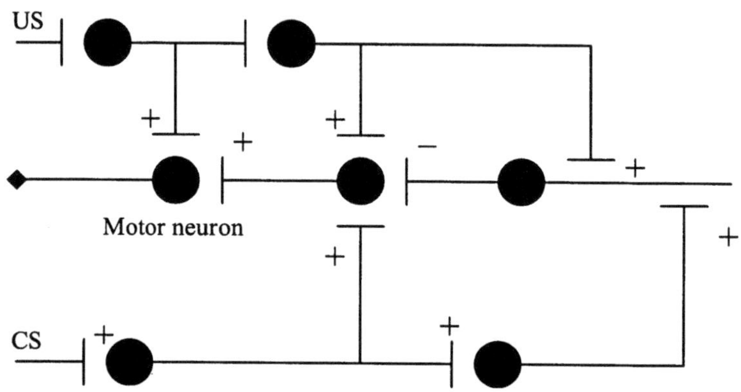

Figure 6. Neural circuit responsible for the eye blink reflex of a rabbit.

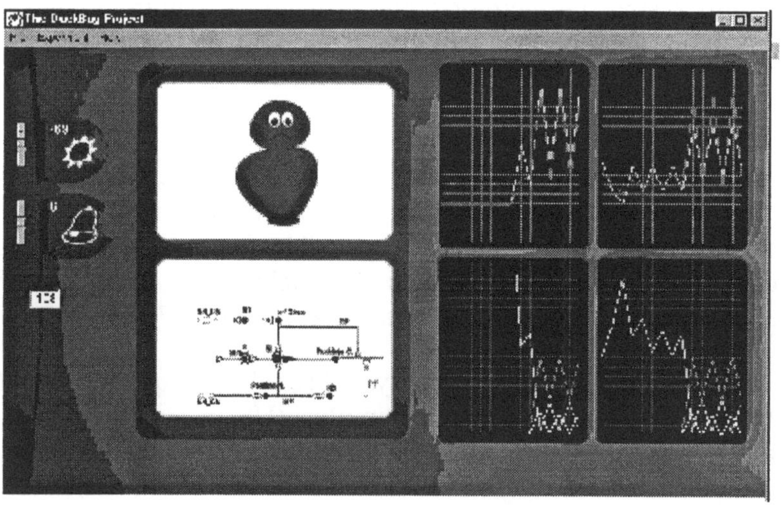

Figure 7. Test environment for (classical and operant) conditioning experiments with Knowbots based on artificial biological neural networks implemented with Java.

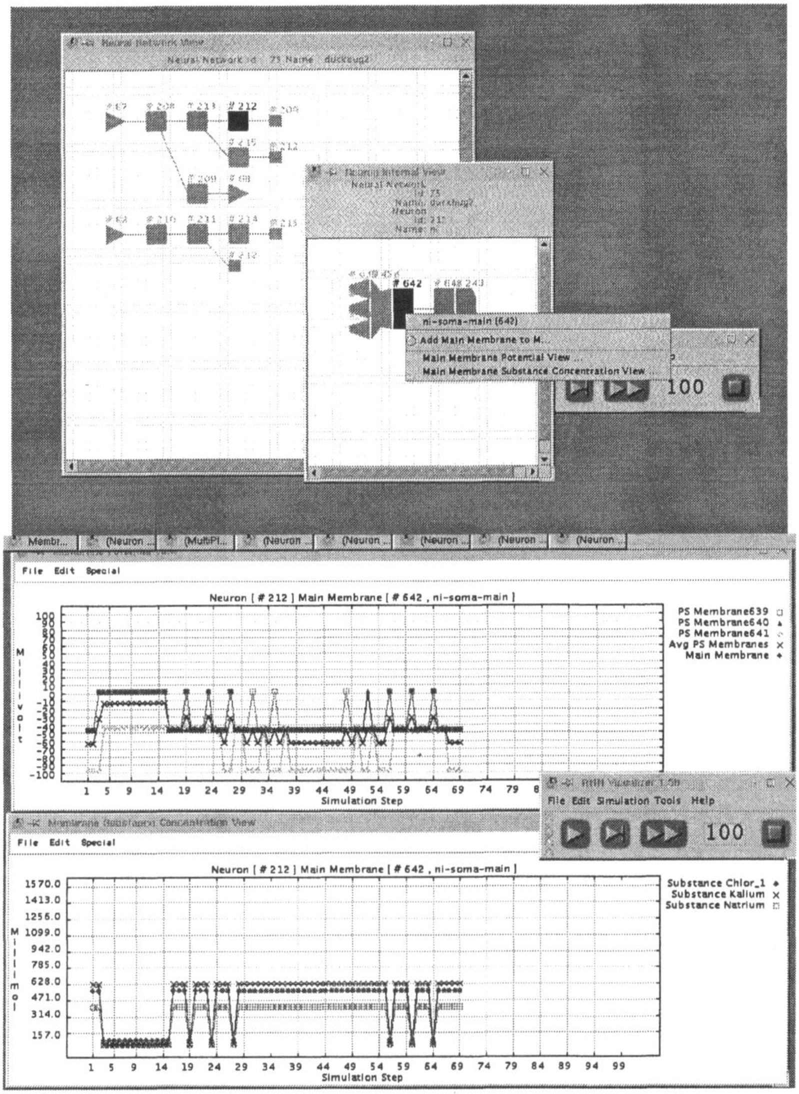

Figure 8. Screenshots of the software brain tools.

The knowbots with embedded BNNs based on our neuron model can deepen our understanding of basic learning behavior and they will be useful to implement intelligent agent software. Knowbotic Systems has developed a set of 'software brain tools' (see Figure 8) combined with 'environment test suits' which enable the user to implement the basic operations of brains that have been identified or suggested so far, such as signal detection, habituation, classical and operant conditioning, orientation and movement in three-dimensional environments.

The neurons and neural networks are described using XML (eXtended Markup Language) and the environments for these networks are based on a multi-agent platform (see 'The artificial environment'). Here follows an example of such an XML description:

```
<neuron id=121 name="n1">
  <soma id=122 name="s">
    <membrane id=123 name="m1">
      <receptor …>
        <neurotransmitter value="GABA"/>
        <neurotransmitter value="Glutamate"/>
        <blocktime value="23"/>
        <event value='false'/>
      </receptor>
      <Na …>
        <pump …>
          <rate value='10'/>
          <threshold value ="245">
          <direction value ="out">
        </pump>
        <rchannel …>
          <rate value='10'/>
          <neurotransmitter value="Glycine"/>
          <blocktime value="40"/>
        </rchannel>
         …
      </Na>
      <K …>
         …
      </K>
      <Cl …>
         …
      </Cl>
    </membrane>
    <hillock id=124 name="h">
     …
    </hillock>
    <LTP1 id=125 …>
       …
    </LTP1>
  </soma>
  <axon ..>
   …
  </axon>
  <bulbend …>
   …
  </bulbend>
</neuron>
```

4 The Artificial Body

Like the artificial intelligence entities in the novel 'Fools War' by Sarah Zettel, the artificial brain needs a body to make it possible to communicate and interact with humans and environments. Robotics' research has shown that intelligent or adaptive behavior is based on a close interaction with the outside world. The measure of learning or intelligence today depends on observable behavior corresponding to well defined learning tasks and environments.

Knowbotic Systems, therefore, concentrates its technical developments on interface technologies, which facilitate the access to knowbots by human users. The most important way to communication is speech. Knowbots are equipped with the speech recognition and synthesis system *AVOICE*. The speaker independent speech recognition is able to identify several different languages. The speech recognition unit can also be trained to understand a specific user to make it possible to handle large dictionaries of several thousands of words. The speech synthesis can read in different formats, such as HTML pages or Word documents. The user has the choice to choose between several 'speakers' with different pronunciation and intonation. AVOICE equips knowbots with a – still limited – human ear and voice. Knowbots connect their users directly to information stored in the Internet and in a company's intranet, regardless of whether the users are using a computer, a telephone, or a mobile phone.

Knowbots can also move around in the Internet, access databases and organize their user dialogues. This is done by *AGENT*, an intelligent search agent and dialogue manager front end. The search agent is able to act as a front end to search robots and crawlers in the World Wide Web. It can also gets access to databases or transform graphical information into text information using standard conversion software utilities encapsulated in an agent framework. Thus, AGENT provides knowbots with a variety of ways to 'perceive' the virtual information universe or infoverse of the Internet (see Figure 9).

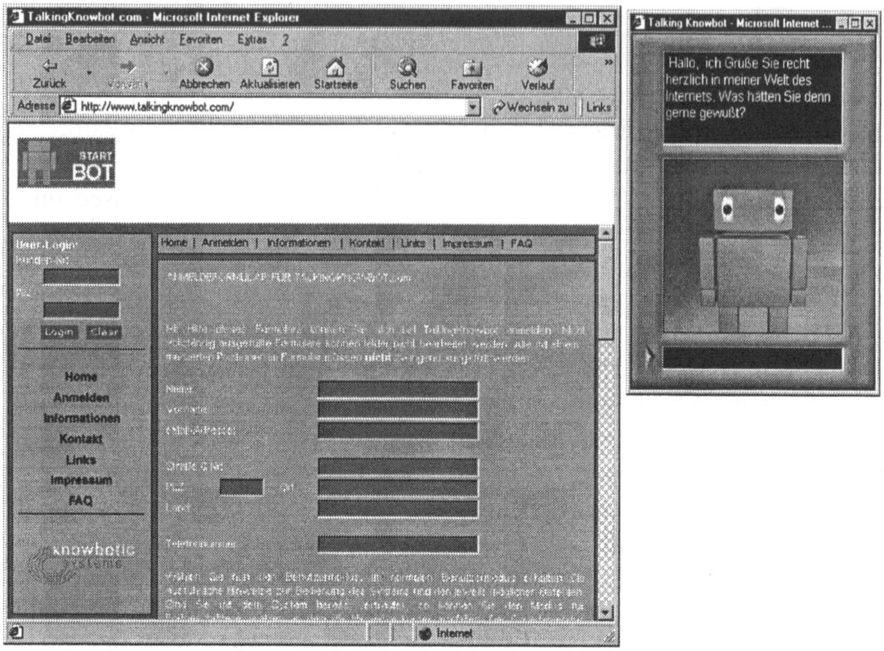

Figure 9. Registration on the AVOICE platform in the Internet (left) – AGENT (right top) is guiding through the registration process and responding to user requests.

5 The Artificial Environment

Today, there is a lot of discussion going on about the irreplaceable value of the human capital and knowledge, but not many actions are taking to maintain and support the development of this capital. Most technical systems concerning the human capital of a company focus on the administration of personnel and training, such as SAP Human Resource modules, Peoplesoft or SABA – just to mention a view of them. An US-American study lists about 300 systems for training administration and delivery. But finding matches of needs and demands in the infoverse certainly means more than matching keywords to indices or user profiles to software agents. The knowledge economy is not so much about information, it is about people. Knowbotic System is, therefore, engaged in a jointly initiative of several partners to implement a Skills Management Information System called 'SMIS' (see Figure 10).

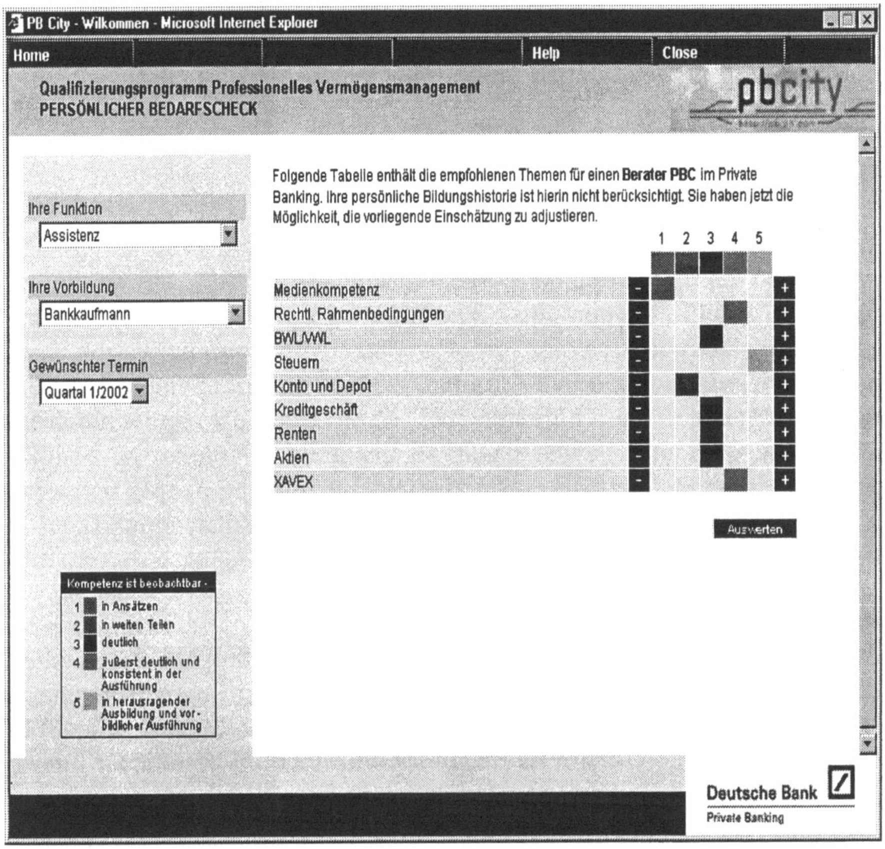

Figure 10. Screen shot from the Skills Management Information System 'SMIS' adapted for Deutsche Bank/Private Banking (Germany) – selection boxes help to identify job roles, prior knowledge, and time schedules (left), a colored table indicates skills covered by recommended training courses (right).

Human users, the users of information systems, visitors and creators of the infoverse, are the main 'component' of a knowbot's environment. Additionally, other knowbots or standardized software agents may also enrich the knowbot environment. For this purpose, Knowbotic Systems has developed one of the first three multi-agent platforms based on FIPA standard (Foundation of Intelligent Physical Agents) standard that is available worldwide. The platform *FATE* (FIPA Agent TEmplate) comprises templates, implemented in JAVA, which allow programmers to convert nearly any computer program into a software agent, that is, the knowbot technology provides easy-to-use ways to introduce a large variety of programs into the virtual learning environment. FATE also allows to run several platforms on different Web sites. This enables knowbots and other agents to communicate, move or replicate themselves all over the World Wide Web.

6 Conclusion

We envision future developments in networked computing and distributed computational intelligence where the users are no longer forced to adapt to the computer. The computers will adapt to the human capabilities to perceive and process data. The communication with computers will adapt to the human way of communication, namely natural language. The computers will be accessible at any time from any point with any device, such as handhelds, laptops, or cellular phones. Computer networks will also extend into the people networks to form the future knowledge network, taking into account specific deficits and potentials of computers and humans.

Knowbots are one of the few holistic visions of a man-machine dialogue in its actual sense, dedicated to support humans where they need help to access and select information - and to learn from them. But knowbots are not the only development in this field. A new level of smart agents and self-learning machines will develop in the near future. Figure 11 summarizes some major developments which are expected in the near future. Among them are software agents, mobile computing, and speech control. But, most of the forecasts of technological growth and development seems to be too conservative.

Up to now, many people are fascinated by the potentials and the exponential growth of the Internet. We do not think that future generations will be too enthusiastic about slow networks, unstructured information heaps and poorly equipped online shopping malls. Smart computers will be part of our every-day life; it will be part of our houses, cars, TV sets, refrigerators, bags and suits. Many ordinary machines and appliances are based on embedded systems, that is, a small specialized computer. Everyday appliances start to become more and more computational things – and they will be smart things in the future. Knowbots and other smart agent technologies will support work, leisure and even cultural or social entertainment. Computers in the form of smart things will make computational intelligence as ordinary as cars or TV sets.

2004	2005	2006	2007	2008	2009	2010
Ubiquitous online learning in universities	Online learning in schools (K12) Interactive communities in the WWW Increasing use of eCommerce	Software agents to search and select information Interactive TV for big audiences	Increasing use of speech recognition and synthesis Central remote control station for 'intelligent buildings'	Increasing use of electronic cash Broadband access to information	Mobile computing and eCommerce	Self-learning software agents 3D virtual reality

Figure 11. Some major developments in interactive media in the next ten years according to a recent study of the Fraunhofer Gesellschaft (Institut für Systemtechnik & Innovationsforschung).

As the computers get closer to their users in future, the people will also get closer to computers: No individual human nor software agent platform will be the master of the infoverse, but partnerships of robots, knowbots, and humans.

References

Bloom, F.E. and Lazerson, A. (1999), *Brain, Mind, and Behaviour* (2nd edition),W.H. Freeman and Company, Worth Publishers.

Doeben-Henisch, G. (1998), "Semiotic machines – an introduction," in Luettich, E.W.B., Mueller, J.E., and Van Zoest, A. (eds.), *Signs & Space,* Raum & Zeichen. An international conference on the semiotics of space and culture, Gunter Narr, Tübingen, pp. 313-327.

Eberhart R., Simpson, P., and Dobbins R. (1996), *Computation Intelligence PC Tools,* Academic Press, New York.

Fogel, D.B. (2000), *Evolutionary Computation, Towards a New Philosophy of Machine Intelligence,* IEEE Press, New York.

Goddard, N. & Hood, N. (1997), "Parallel GENESIS for large scale modeling," in Bower, J.M. (ed.), *Computational Neuroscience '96*, Plenum, New York.

Harasim, L.M., Hiltz, S.R., Teles, L., and Turoff, M. (eds.) (1995), *Learning Networks: a Field Guide to Teaching and Learning Online*, MIT Press, Boston.

Hasebrook, J. (1995), "Lernen mit Multimedia (Learning with multimedia)," *German Journal of Educational Psychology*, vol. 9, no. 2, pp. 95-103.

Hasebrook, J. and Nathusius, W. (1997), "An expert advisor for vocational guidance," *Journal of Artificial Intelligence in Education,* vol. 8, no. 1, pp. 21-41.

Hasebrook, J.P. (1998), "Searching the Web without losing the mind," *Webnet Journal*, vol. 1, no.2, pp.24-32

Haykin, S. (1994), *Neural Networks, a Comprehensive Foundation*, New York: Macmillan College Publishing Company.

Kosko, B. (1992), *Neural Networks for Signal Processing*, Prentice-Hall, Englewood Cliffs, NJ.

Maass, W. (1997), "Networks of Spiking Neurons: the Third Generation of Neural Network Models," Graz: Institute for Theoretical Computer Science, TU Graz.

McDermott, D. and Poggio, T. (1989), "Artificial intelligence," in Hopcroft, J.E. and Kennedy, K.W. (chairs), *Computer Science Achievements and Opportunities*, Report of the NSF Advisory Committee for Computer Research, Society for Industrial and Applied Mathematics, Philadelphia, pp. 41-50.

Menzel, R. (1996), "Neuronale Plastizität, Lernen und Gedächtnis," in Dudel, J., Menzel, R., and Schmidt, R.F. (eds.), *Neurowissenschaft. Vom Molekül zur Kognition*, Springer, New York, pp. 485-518.

Moravec, H. (1998), "When will computer hardware match the human brain?" *Journal of Transhumanism*, vol. 1, http://www.trans humanist.com/volume1/moravec.htm .

Sheperd, G.M. (ed.) (1998), *The Synaptic Organization of the Brain*, Oxford University Press, Oxford.

Sproull, L. and Kiesler, S. (1991), *Connections: New Ways of Working in the Networked Organization*, MIT Press, Cambridge MA.

Turing, A.M. (1950), "Computing machinery and intelligence," *Mind*, vol. 59, pp. 433-460.

Chapter 4

Ontologies in Web Intelligence

N. Zhong

Ontologies and agent technology can play a crucial role in enabling Web-based knowledge processing, sharing, and reuse between applications. The chapter investigates the roles of ontologies in Web Intelligence. Three ontology categories are suggested, some of the research and development with respect to the three categories is presented, the major ontology languages are surveyed, and a multi-phase process of automatic construction of the *domain-specific* ontologies is discussed.

1 Introduction

With the rapid growth of Internet and World Wide Web (WWW), we have now entered into a new information age. The Web has significant impacts on academic research, business, and our everyday life. It revolutionizes the way in which information is gathered, stored, processed, and used. The Web offers new opportunities and challenges for many areas, such as business, commerce, marketing, finance, publishing, education, research and development.

Web Intelligence (WI) exploits Artificial Intelligence (AI) and advanced Information Technology (IT) on the Web and Internet. It is the key and the most important research field of IT for business intelligence. WI emphasizes to study all aspects of Web-based intelligent information technology systematically, such as Web information system environment and foundations, Web human-media engineering, Web information management, Web information retrieval, Web agents, Web mining and farming, and Web-based applications (Zhong *et al.*, 2000).

Ontologies and agent technology can play a crucial role in Web intelligence by enabling Web-based knowledge processing, sharing, and reuse between applications. Generally defined as shared formal conceptualizations of particular domains, ontologies provide a common understanding of topics that can be communicated between people and agent-based systems.

The chapter investigates the roles of ontologies in Web Intelligence. In Section 2, representation of ontologies is discussed, three ontology categories are suggested, and some of the research and development with respect to the three categories is presented. In Section 3, the roles of ontologies for Web Intelligence are described, and the major ontology languages for Web intelligence are surveyed. In Section 4, a multi-phase process of automatic construction of the *domain-specific* ontology is discussed. Finally, Section 5 gives concluding remarks.

2 Representation and Categories of Ontologies

An ontology is a specification a conceptualization. Ontologies are content theories about the sorts of objects, properties of objects, and relations between objects that are possible in a specified domain of knowledge (Chandrasekaran *et al.*, 1999).

An ontology typically contains a network of concepts within a domain and describes each concept's crucial properties through an attribute-value mechanism. Such network is either directed or undirected one. It might also be a special type of network, that is, a concept hierarchy (tree). Further relations between concepts might be described through additional logical sentences.

An ontology can be very high-level, consisting of concepts that organize the upper parts of a knowledge base, or it can be domain-specific such as a chemical ontology. We here suggest three categories of ontologies: *domain-specific*, *task*, and *universal* ones.

A *domain-specific* ontology describes a well-defined technical or business domain.

A *task* ontology might be either a quite domain-specific one, or a set of ontologies with respect to several domains (or their reconstruction for

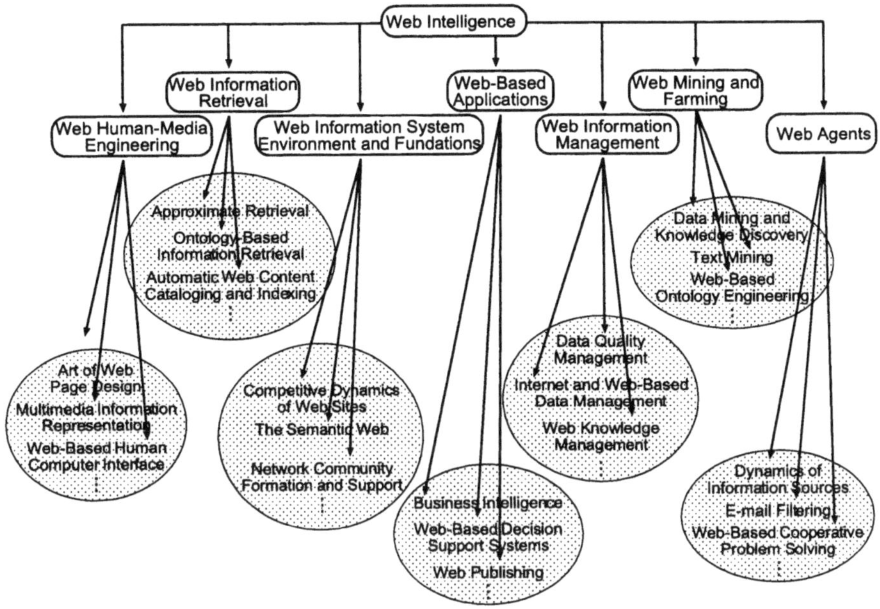

Figure 1. A domain-specific ontology on WI (Web Intelligence).

that task), in which relations between ontologies are described for meeting the requirement of that task.

An *universal* ontology describes knowledge at higher levels of generality. It is a more general-purpose ontology (or called a common ontology) that is generated from several domain-specific ontologies. It can serve as a bridge for communication among several domains or tasks.

Figures 1 and 2 show two domain-specific ontologies on WI and IAT (Intelligent Agent Technology), respectively, which are presented by a concept hierarchy. Figure 3 shows a task ontology that a set of ontologies with respect to WI and IAT ontologies. Figure 4 shows a task ontology on software marketing, which is presented by a directed concept network. In Section 4, we will describe a way of automatic construction of domain-specific ontologies presented by a directed concept network.

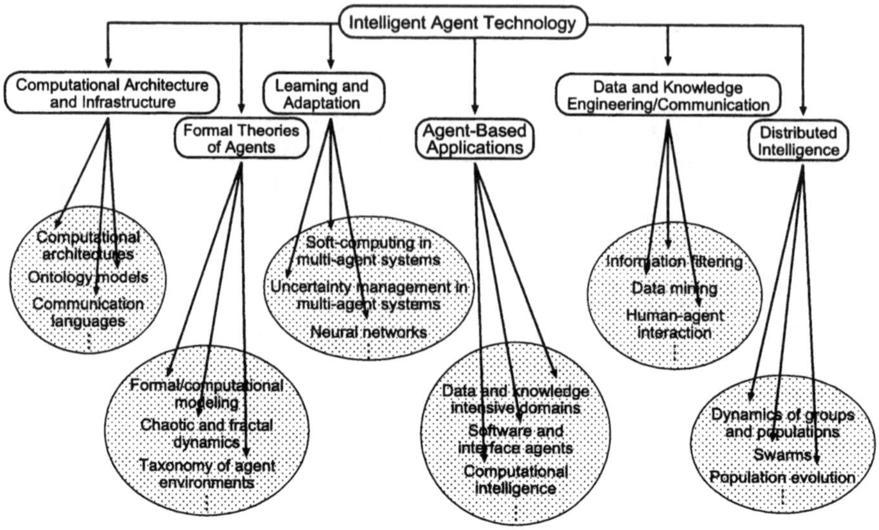

Figure 2. A domain-specific ontology on IAT (Intelligent Agent Technology).

3 Ontologies for Web Intelligence

This section discusses the roles of ontologies and ontology languages for Web intelligence.

3.1 The Roles of Ontologies

Generally speaking, a domain-specific (or task) ontology forms the heart of any knowledge information system for that domain (or task). Ontologies will play a major role in supporting information exchange processes in various areas. The roles of ontologies for Web intelligence include:

- communication between Web communities,
- agents communication based on semantics,
- knowledge-based Web retrieval,
- understanding Web contents in a semantic way,
- and Web community discovery.

More specifically, new requirements for any exchange format on the Web are:

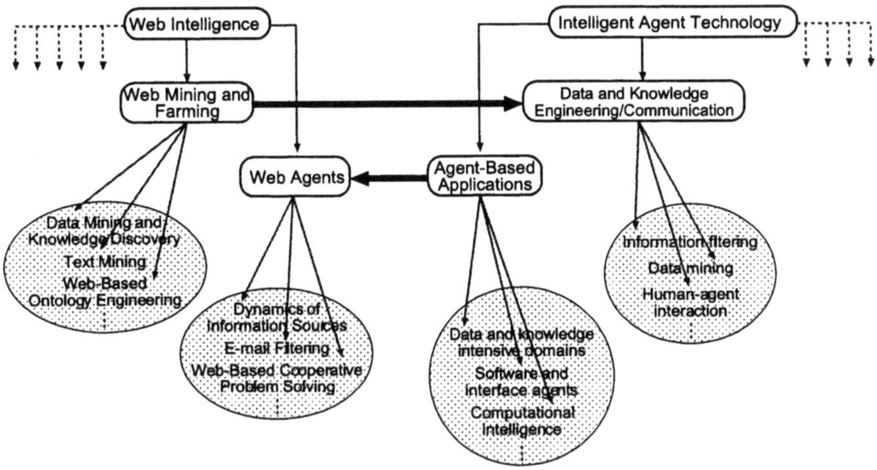

Figure 3. A task ontology.

- Universal expressive power.

  A Web based exchange format must be able to express any form of data.

- Syntactic interoperability.

  Applications must be able to read the data and get a representation that can be exploited.

- Semantic interoperability.

  One of the most important requirements for an exchange format is that data must be understandable. It is about defining mappings between terms within the data, which requires content analysis.

One of the fundamental issues of WI is to study the *semantics* in the Web, called the "semantic Web", that is, modeling *semantics* of Web information. Advantages of the semantic Web include:

- Allowing more of the Web content (not just form) to become machine readable and processible,
- Allowing for recognition of the semantic context in which Web materials are used,
- Allowing for the reconciliation of terminological differences between diverse user communities.

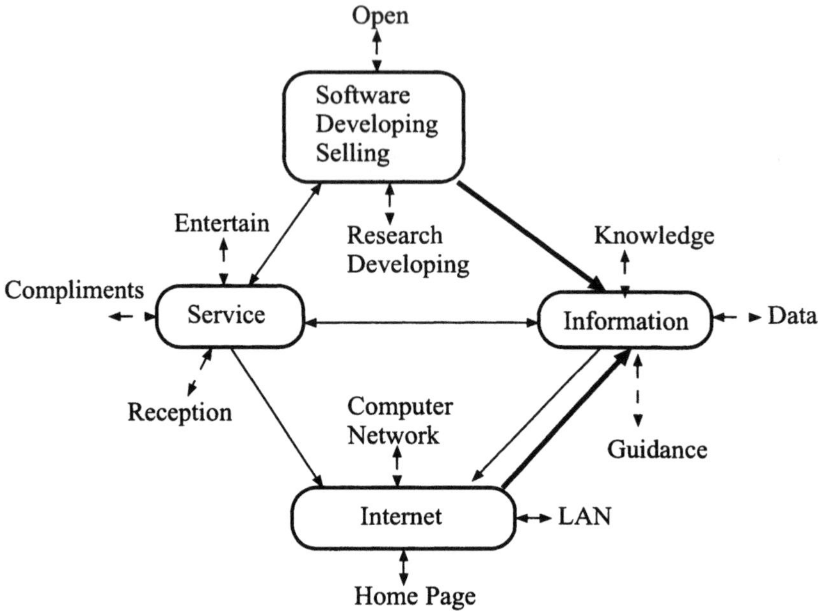

Figure 4. A task ontology on software marketing.

Thus, information will be machine-processible in ways that support intelligent network services such as information brokers and search agents (Decker *et al.*, 2000).

The semantic Web requires interoperability standards that address not only the syntactic form of documents but also the semantic content. A semantic Web also lets agents utilize all the data on all Web pages, allowing it to gain knowledge from one site and apply it to logical mappings on other sites for ontology-based Web retrieval and e-business intelligence (Martin and Eklund, 2000). For instance, ontologies can be used in e-commerce to enable machine-based communication between buyers and sellers, vertical integration of markets, description reuse between different marketplaces. Web-search agents use ontologies to find pages with words that are syntactically different but semantically similar.

In summary, ontologies and agent technology can play a crucial role in enabling such Web-based knowledge processing, sharing, and reuse between applications.

3.2 Ontology Languages

Ontologies provide a way of capturing a shared understanding of terms that can be used by human and programs to aid in information exchange. Ontologies have been gaining popularity as a method of providing a specification of a controlled vocabulary. Although simple knowledge representation such as Yahoo's taxonomy provides notions of generality and term relations, classical ontologies attempt to capture precise meanings of terms. In order to specify meanings, an ontology language must be used. So far, several ontology languages such as OIL, SHOE, and DAML have been proposed.

OIL (Ontology Inference Layer) is an Ontology Interchange Language for the Web (Fensel *et al.*, 2000; Decker *et al.*, 2000). It is an effort to produce a layered architecture for specifying ontologies. The major functions of OIL include:

- It provides the modeling primitives commonly used in frame-based ontologies.
- It has a simple, clean, and well defined semantics based on description logics.
- Automated reasoning support may be specified and provided in a computationally efficient manner.

SHOE (Simple HTML Ontology Extensions) is an extension to HTML which provides a way to incorporate machine-readable semantic knowledge in HTML or other World-Wide Web documents such as XML (Heflin and Hendler, 2000; Luke *et al.*, 1997). It provides:

- A hierarchical classification mechanism for HTML documents (and optionally non-HTML documents) or subsections of HTML documents.
- A mechanism for specifying relationships between classified elements and other classified elements or specific kinds of data (numbers, dates, etc.)
- A simple way to specify ontologies containing rules that define valid classifications, relationships, and inferred rules.

DAML (DARPA Agent Markup Languages) is a new DARPA research program. One of main tasks of this program is to create an Agent Mark-Up Language (DAML) built upon XML. It is a *semantic* language that ties the information on a page to machine readable semantics (ontology). It is a step toward a "semantic Web" where agents, search engines and other programs can read DAML mark-up to decipher meaning rather than just content on a Web site (Hendler, 2000).

4 Automatic Construction of Ontologies

Automatic construction of ontologies is a challenge task in both ontology engineering and WI. This section describes a process of construction of task (or domain-specific) ontologies. This process is a multi-phase process in which various text mining and natural-language understanding methods are used (Zhong *et al.*, 2000; Zhong, 2001).

Much of data is now in textual form. This could be data on the Web, e-mails, e-library, or electronic papers and books, among others, namely *text databases* in this chapter. Text mining is to mine knowledge from semi-structured or unstructured text. Text mining is a multidisciplinary field, involving various techniques such as data mining, information retrieval, natural-language understanding, case-based reasoning, statistics, and intelligent agent technology.

Figure 5 shows a sample process of construction of a task ontology on software marketing. The major steps in the process include morphological analysis, text classification, generation of classification rules, conceptual relationship analysis, generation of ontology, as well as refinement and management of ontology. A thesaurus is necessary to be used as a background knowledge base in the process. We emphasize that the process is iterative, and may repeat at different intervals when new/updated data come. We discuss the major techniques used in the process and show some preliminary results.

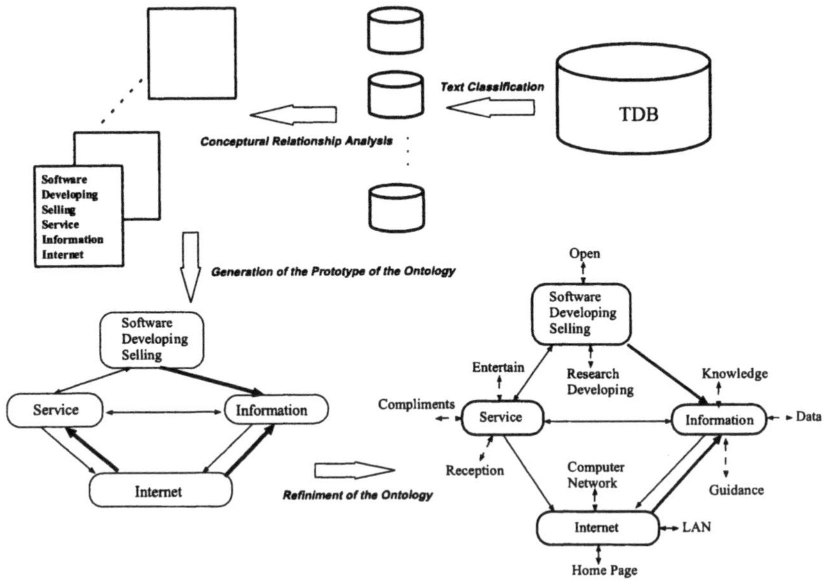

Figure 5. A sample process of construction of the ontology.

4.1 Text Classification

In order to discover a task (or domain-specific) ontology from text databases, we first need to annotate the texts with class labels. This annotation task is that of text classification. However, it is expensive that the large amounts of texts are manually labeled. This section introduces a semi-automatic approach to classify text databases, which is based on uncertainty sampling and probabilistic classifier. The main contribution of ours is to extend the method proposed by Lewis (Lewis and Catlett, 1994) for multiple classes classification.

We use a variant of the Bayes' rule below:

$$P(C|w) = \frac{\exp(\log \frac{P(C)}{1-P(C)} + \sum_{i=1}^{d} \log(P(w_i|C)/P(w_i|\overline{C})))}{1 + \exp(\log \frac{P(C)}{1-P(C)} + \sum_{i=1}^{d} \log(P(w_i|C)/P(w_i|\overline{C})))} \quad (1)$$

where $w = \{w_1, \ldots, w_d\}$ is a set of the terms in a text, and C is a class. Although we treat, in this equation, only two classes $C_1 = C$ and $C_2 = \overline{C}$ with $P(\overline{C}) = 1 - P(C)$, it can be extended to deal with multiple classes classification by using the method to be stated in the end of this section.

However, Eq. (1) is rarely used directly in text classification, probably because its estimates of $P(C|w)$ are systematically inaccurate. Hence we use Logistic regression, which is a general technique for combining multiple predictor values to estimate a posterior probability, in Eq. (1). Thus, we obtain the following equation:

$$P(C|w) = \frac{\exp(a + b\sum_{i=1}^{d} \log(P(w_i|C)/P(w_i|\overline{C})))}{1 + \exp(a + b\sum_{i=1}^{d} \log(P(w_i|C)/P(w_i|\overline{C})))}. \quad (2)$$

Intuitively, we could hope that the logistic parameter a would substitute for the hard-to-estimate prior log odds in Eq. (1), while b would serve to dampen extreme log likelihood ratios resulting from independence violations.

Furthermore, we use the following equation to estimate the values $P(w_i|C)P(w_i|\overline{C})/P(w_i|C)P(w_i|\overline{C})$ as the first step in using Eq. (2),

$$\frac{P(w_i|C)}{P(w_i|\overline{C})} = \frac{\frac{c_{pi}+(N_p+0.5)/(N_p+N_n+1)}{N_p+d(N_p+0.5)/(N_p+N_n+1)}}{\frac{c_{ni}+(N_n+0.5)/(N_p+N_n+1)}{N_n+d(N_n+0.5)/(N_p+N_n+1)}} \quad (3)$$

where N_p and N_n are the numbers of terms in the positive and negative training sets, respectively, c_{pi} and c_{ni} are correspondingly the numbers of examples of w_i in the positive and negative training sets, respectively, and d is the number of different terms in a text.

Based on the preparation stated above, we briefly describe the main steps of text classification below:

Step 1. Select examples (terms) as an initial classifier for N classes by a user and all the N classes are regarded as a set of the negative classes.

Step 2. Select a class from the set of the negative classes as a positive class, and the remaining ones are regarded as a set of the negative classes.

Step 3. While a user is willing to label texts.

Step 3.1 Apply the current classifier to each unlabeled text.

Step 3.2 Find the k texts for which the classifier is least certain of class membership by computing their posterior probabilities in Eq (2).

Step 3.3 Have the user label the subsample of k texts.

Step 3.4 Train a new classifier on all labeled texts.

Step 4. Repeat *Step 2* to *Step 3* until all classes were selected as a positive class.

Selecting examples (terms) as an initial classifier by a user is an important step because of the need for personalization applications. The requirements and biases of a user are represented in the classifier.

For example, we have a text database in which there are a lot of mixed texts on soccer teams, software marketing, hot-spring, etc. This database has been pre-processed by using morphological analysis. Thus we may use the text classification method stated above to obtain the classified sub-databases on soccer teams, software marketing, hot-spring, respectively.

4.2 Generation of Ontology

Based on the result of text classification, the process of generation of ontology can be divided into the following two major stages.

The first stage is *conceptual relationship analysis* (Salton, 1989; Chen and Lynch, 1992). We first compute the combined weights of terms in texts by Eqs. (4) and (5), respectively.

$$D_i = \log d_i \times tf_i \tag{4}$$

$$D_{ij} = \log d_{ij} \times tf_{ij} \tag{5}$$

where d_i and d_{ij} are the text frequency, which represent the numbers of texts in a collection of n texts in which term i occurs, and both term i and term j occur, respectively. tf_i and tf_{ij} are the term frequencies, which represent the numbers of occurrences of term i, and both term i and term j, in a text, respectively.

Then a network-like concept space is generated by using the following equations to compute their similarity relationships.

$$Rel(i,j) = \frac{D_{ij}}{D_i} \tag{6}$$

Table 1. The similarity relationships of the terms.

Term i	Term j	$Rel(i,j)$
team	soccer	0.7385
league	soccer	0.7326
university	soccer	0.5409
player	soccer	0.4929
Japan	soccer	0.4033
region	soccer	0.4636
game	soccer	0.1903
sports	soccer	0.1803
gymkhana	soccer	0.1786
soccer	team	0.7438
league	team	0.8643
university	team	0.5039
player	team	0.1891
Japan	team	0.1854
region	team	0.1973
...	...	...

$$Rel(j,i) = \frac{D_{ji}}{D_j}. \tag{7}$$

Here Eq. (6) and Eq. (7) compute the relationships from term i to term j, and from term j to term i, respectively. We also use a threshold value to ensure that only the most relevant terms are remained. Table 1 shows a portion of the similarity relationships of the terms on soccer teams.

The second stage is to generate the prototype of the ontology by using a variant of the Hopfield network. Each remaining term is used as a neuron (unit), the similarity relationship between term i and term j is taken as the unidirectional and weighted connection between neurons. At time 0,

$$\mu_i(0) = x_i : 0 \leq i \leq n-1$$

where $\mu_i(t)$ is the output of unit i at time t, and x_i indicates the input pattern with a value between 0 and 1. At time 0, only one term receive the value 1 and all other terms receive 0. We repeat to use the following equation n times (i.e. for n terms):

$$\mu_i(t+1) = f_s[\sum_{i=0}^{n-1} t_{ij}\mu_i(t)], \quad 0 \leq j \leq n-1 \tag{8}$$

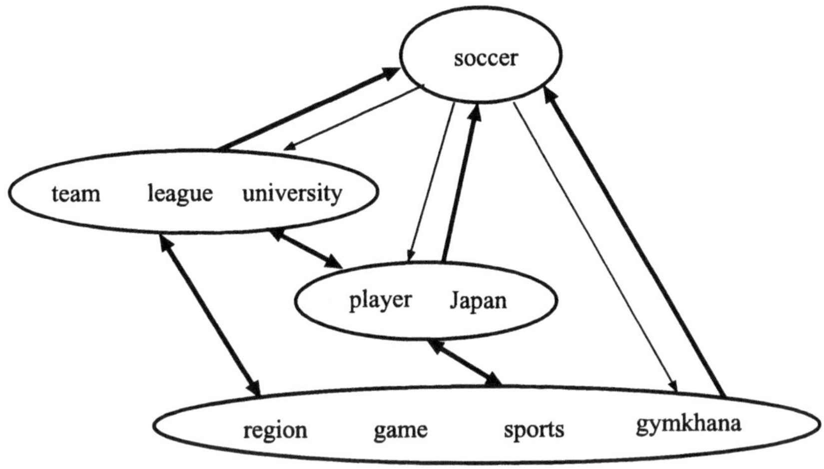

Figure 6. The prototype of a task ontology on soccer teams.

where t_{ij} represents the similarity relationship $Rel(i, j)$ as shown in Eq.(6) (or Eq.(7) for t_{ji}), f_s is the sigmoid function as shown below:

$$f_s(net_j) = \frac{1}{1 + exp[(\theta_j - net_j)/\theta_0]} \tag{9}$$

where $net_j = \sum_{i=0}^{n-1} t_{ij}\mu_i(t)$, θ_j serves as a threshold or bias, and θ_0 is used to modify the shape of the sigmoid function.

This process is repeated until there is no change between two iterations in terms of output, that is, it converged by checking the following equation:

$$\sum_{j=0}^{n-1}[\mu_j(t+1) - \mu_j(t)]^2 \leq \varepsilon \tag{10}$$

where ε is the maximal allowable error.

The final output represents the set of terms relevant to the starting term, which can be regarded as the prototype of a task (or domain-specific) ontology. Figure 6 shows an example of the prototype of a task ontology on soccer teams. It is generated by using each term shown in Table 1 as a starting input pattern for learning on the Hopfield network.

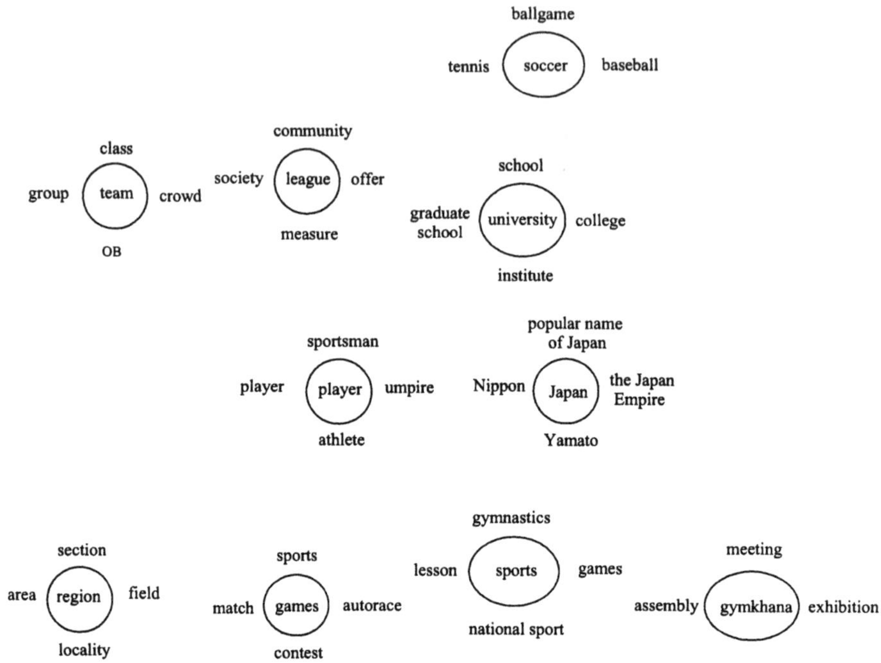

Figure 7. A task ontology refined by using thesaurus.

4.3 Refinement of Ontology

There is often a limit to the construction of ontology from text databases, whatever the technique employed. Incorporating any associated knowledge significantly increases the efficiency of the process and the quality of the ontology generated from the text data. A thesaurus is a useful source to be used as a background knowledge base for refinement of ontology. By using the thesaurus, the terms are extended by including their synonym, wider and narrow sense of the terms. Figure 7 shows a task ontology on soccer teams refined by using thesaurus. Note the concept space shown in Figure 7 is the same as the one shown in Figure 6, but we omitted the links between terms for clarity of the figure.

5 Conclusions

The chapter presented the roles of ontologies for Web intelligence, including issues on presentation, languages, and automatic construction of ontologies. A task (or domain-specific) ontology forms the heart of any knowledge information system for that task (or domain). Ontologies will play a major role in supporting information exchange processes in various areas. On the other hand, agent technology is required since information on the Web is distributed. The integration of ontologies and agent technology increases the autonomy of Web-based information systems.

We emphasize that the process of automatic construction of ontologies is iterative, and may repeat at different intervals when new/updated data come. Hence how to handle change is an important issue related to refinement of ontology. In particular, during the (long) lifetime of an application session, there may be many kinds of changes such as changes in the text data, the purpose of using both the text data and the ontology, etc. Hence we need a method to reuse the exiting ontology with local adjustment adapted to the changes.

Acknowledgment

This work was partially supported by Telecommunications Advancement Foundation (TAF).

References

Aggarwal, C.C. and Yu, P.S. (1999), "On Text Mining Techniques for Personalization", Zhong, N., Skowron, A., and Ohsuga, S. (eds.) *New Directions in Rough Sets, Data Mining, and Granular-Soft Computing*, LNAI 1711, Springer-Verlag, pp. 12-18.

Chandrasekaran, B., Josephson, J.R., and Benjamins, V.R. (1999), "What Are Ontologies, and Why Do We Need Them?", *IEEE Intelligent Systems*, Vol. 14, No. 1, pp. 20-26.

Chen, H. and Lynch, K.J. (1992), "Automatic Construction of Networks of Concepts Characterizing Document Databases", *IEEE Tran. on*

Sys. Man and Cybernetics, Vol. 22, No. 5, pp. 885-902.

Chen, H. (1994), "Collaborative Systems: Solving the Vocabulary Problem", *IEEE Computer*, Vol. 27, No. 5, pp. 58-66.

Cooper, W.S., Gey, F.C., and Dabney. D.P. (1992), "Probabilistic Retrieval Based on Staged Logistic Regression", *Proc. ACM SIGIR'92*, pp. 198-210.

Cooley, R., Mobasher, B., and Srivastavva, J. (1999), "Data Preparation for Mining Would Wide Web Browsing Patterns", *Knowledge and Information Systems, An International Journal*, Vol. 1, No. 1, Springer-Verlag, pp. 5-32.

Decker, S., Melnik, S. et al. (2000), "The Semantic Web: The Roles of XML and RDF", *IEEE Internet Computing*, Vol. 4, No. 5, pp. 63-74.

Decker, S., Mitra, P., and Melnik, S. (2000), "Framework for the Semantic Web: An RDF Tutorial", *IEEE Internet Computing*, Vol. 4, No. 6, pp. 68-73.

Fensel, D. et al. (2000), "OIL in a Nutshell", R. Dieng and O. Corby (eds.) *Knowledge Engineering and Knowledge Management: Methods, Models, and Tools*, LNAI 1937, Springer-Verlag, pp. 1-16.

Frank, G., Farquhar, A., and Fikes, R. (1999), "Building a Large Knowledge Base from a Structured Source", *IEEE Intelligent Systems*, Vol. 14, No. 1, pp. 47-54.

Guarino, N. (ed.) (1998), *Formal Ontology in Information Systems*, IOS Press.

Heflin, J. and Hendler, J. (2000), "Dynamic Ontologies on the Web", *Proc. AAAI-2000*, AAAI/MIT Press, pp. 443-449.

Hendler, J.A. (2000), "Agent Based Computing and DAML", http://www.daml.org/.

Lewis, D.D. and Catlett, J. (1994), "Heterogeneous Uncertainty Sampling for Supervised Learning", *Proc. Eleventh Inter. Conf. on Machind Learning*, pp. 148-156.

Liu, J. and Zhong, N. (eds.) (1999), *INTELLIGENT AGENT TECHNOLOGY: Systems, Methodologies, and Tools*, World Scientific.

Liu, J., Zhong, N., Tang, Y.Y., and Patrick S.P. Wang (eds.) (2001), *Agent Engineering*, World Scientific.

Luke, S., Spector, L., Rager, D., and Hendler, J. (1997), "Ontology-based Web Agents", *Proc. First International Conference on Autonomous Agents (Agents97)*, ACM Press, pp. 59-66.

Martin, P. and Eklund, P.W. (2000), "Knowledge Retrieval and the World Wide Web", *IEEE Intelligent Systems*, Vol. 15, No. 3, pp. 18-25.

Salton, G. (1989), *Automatic Text Processing*, Addison-Wesley Publishing.

Yao, Y.Y. and Zhong, N. (1999), "An Analysis of Quantitative Measures Associated with Rules", Zhong, N. and Zhou, L. (eds.) *Methodologies for Knowledge Discovery and Data Mining*, LNAI 1574, Springer-Verlag, pp. 479-488.

Zhong, N., Kakemoto, Y., and Ohsuga, S. (1997), "An Organized Society of Autonomous Knowledge Discovery Agents", Peter Kandzia and Matthias Klusch (eds.) *Cooperative Information Agents*, LNAI 1202, Springer-Verlag, pp. 183-194.

Zhong, N., Yao, Y.Y., and Kakemoto, Y. (2000), "Automatic Construction of Ontology from Text Databases", N. Ebecken and C.A. Brebbia (eds.) *Data Mining*, Vol. 2, WIT Press, pp. 173-180.

Zhong, N., Liu, J., Yao, Y.Y. and Ohsuga, S. (2000), "Web Intelligence (WI)", Proc. the 24th IEEE Computer Society International Computer Software and Applications Conference (COMPSAC 2000), a position paper for a panel on Data Mining and Web Information Systems, pp. 469-470.

Zhong, N. (2001), *Knowledge Discovery and Data Mining*, in the Encyclopedia of Microcomputers, Vol. 27 (Supplement 6) Marcel Dekker, pp. 235-285.

Zhong, N. and Ohsuga, S. (2001), *Automatic Knowledge Discovery in Larger Scale Knowledge-Data Bases*, in C. Leondes (ed.) The Handbook of Expert Systems, Vol.4, Academic Press, pp. 1015-1070.

Chapter 5

Software Agents for Internet-Based Systems and Their Design

H.H. Pham

Nowadays many business and engineering systems are Internet-based. This chapter describes how software agents can be used to build autonomous and collaborative Internet-based systems and how agents themselves participate in developing and designing these systems. We will introduce the principles of Internet agent-based systems, their classifications, architectures, design procedure, and applications.

1 Introduction

The Internet has changed the way business is done. Not only marketing, cataloging, and trading can be conducted via the Internet (Bakos 1998, Beck 1999, Huhns 1999, Froehlich 1999) but also manufacturing and production management, monitoring and controls of engineering processes can take advantages of Internet environment (Elofson 1998, Evans1997, McFall1998).

Internet-based systems can carry out business, manufacturing, and engineering activities in *a sharable, interactive, distributed and scalable* fashion. Information in these systems can be created, accessed and shared among many parties regardless of their locations. Users can interact with business or engineering processes via *user-friendly interfaces*. Internet-based systems are also flexible. They can serve a wide number of users and may be self-extending without the need for rebuilding the systems.

It is difficult to integrate a number of Internet-based systems. While each Internet-based system is interactive with the human users it is *not interactive with the other Internet-based systems*. Besides, as the initial and primary purpose of Internet-based systems is to work with human users they play a secondary role in solving a problem and are *not autonomous*. They need specific commands from human users and return replies to the users in a client-server manner. They are a supporting component and a source of information rather than a problem solver.

These limitations can be overcome by using relevant software agents (Khare 1997, Bradshaw 1999, Ferber 1999). The integration with agents makes Internet-based systems more autonomous and collaborative (Arisha 1999, Baker 1999, Shen1999, Grecu 1996). When components of these systems are agent-based they can interact with each other by providing communication between different systems. This is done by exchanging data through agent communication or by exchanging codes through agent migration. Inheriting agent concepts and technologies, Internet agent-based systems can work independently from human users and can accomplish complete tasks as problem solvers. Thus, Internet-based systems using software agents would be a powerful and suitable means to efficiently manage distributed business and engineering systems.

This chapter provides an overview of Internet agent-based systems. It introduces technologies for building Internet agents and analyzes how they can be integrated and used in Internet-based systems and how agents themselves participate in developing and designing these systems.

The rest of this chapter is organized as follows. In Section 2, we study the principles of Internet agent-based systems (IAS) as an integration of distributed computing and artificial intelligence. Then, we classify IAS based on different criteria in Section 3. In Section 4, we discuss communication and coordination issues in IAS. Next, we describe the main steps in designing IAS. In Section 6, we explain how agents could be used for designing IAS. We use two case studies in Section 7 to show the effectiveness and practical usage of agents in IAS. The conclusions are given in Section 8.

2 Principles of Internet Agent-Based Systems

Internet agent-based systems (IAS) is an integration of distributed computing and artificial intelligence. We will introduce the principles of IAS where distributed computing techniques are used to construct the "bodies" of IAS while artificial intelligence methodologies are the foundations of their "minds". Internet agent-based systems are based on computing technologies of two fields in computer science: (i) distributed computing, for the Internet features; and (ii) artificial intelligence, for the agents features. We first investigate them separately and then in combination.

2.1 Internet-Based Components

Internet-based systems use client-server mechanisms with a server site and a dynamic number of client sites as shown in Figure1. Their components are divided into: (i) server components; and (ii) client components. Server components are stored and run at server site and are supported by a Web server. Client components are stored at the server site but run at the client sites supported by Web browsers.

A scenario of how Internet-based clients and server components work is shown in Figure 1. A user, through a web browser, which is the client supporting environment, makes the first request from the client site to the server site by specifying a web address. When this request reaches the server site it goes through the server supporting environment or web server which then looks for the required object. In this case, a html file, and returns a copy of it to the given client site where it will execute. Thus, replications of html files are supported. When there are many clients each requesting the same html file, each of them receives their own copy. That is how the Internet-based systems are scalable. They can be extended automatically without rebuilding any part of the system. Meanwhile, they have minimal usage of system resources as a server component is invoked only when there is a request from client site.

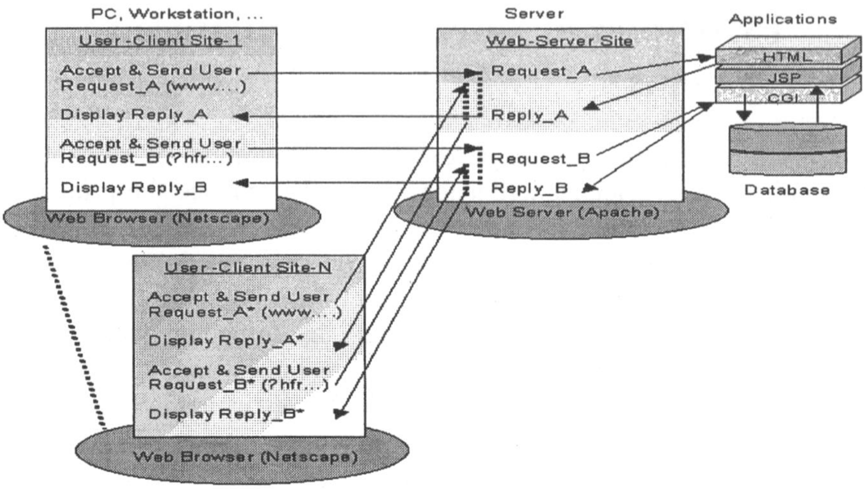

Figure 1. Internet-based components.

2.2 Agent-Based Components

Software agents are programs that can run and act autonomously without human user supervision (Knapik 1998, Moses 1999). A software agent which works on the Internet or Intranets can have: (i)sensors; (ii) thinking engine; (iii) knowledge and data bases; (iv) actuators; and (v) Internet engine. The organization of Internet agents is shown in Figure 2.

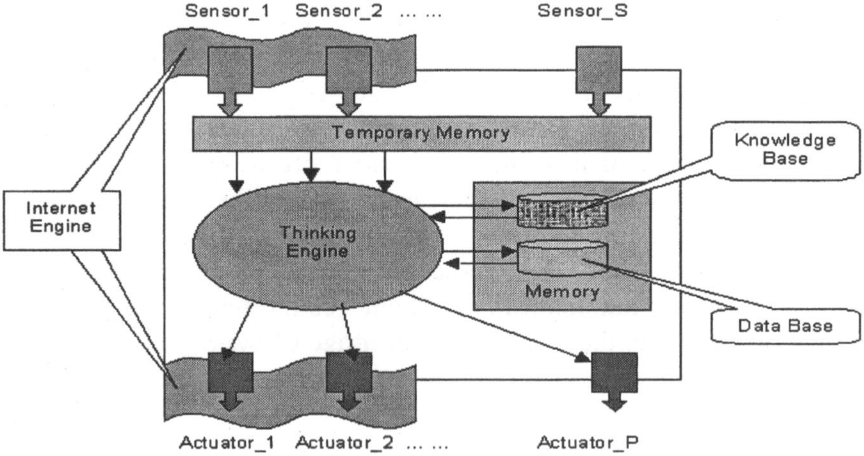

Figure 2. Structure of an Internet agent.

An agent-based component may employ one, several, or a dynamic number of agents which interact with each other to implement user tasks. While agents are managed at logical level by a multi-agent system (MAS) using AI techniques, their actions are carried out at physical level as regular programs using technologies of distributed computing. Agent interactions are carried out via agent communication. A mapping of the logical representation to the physical management for agent activities is shown in Figure 3.

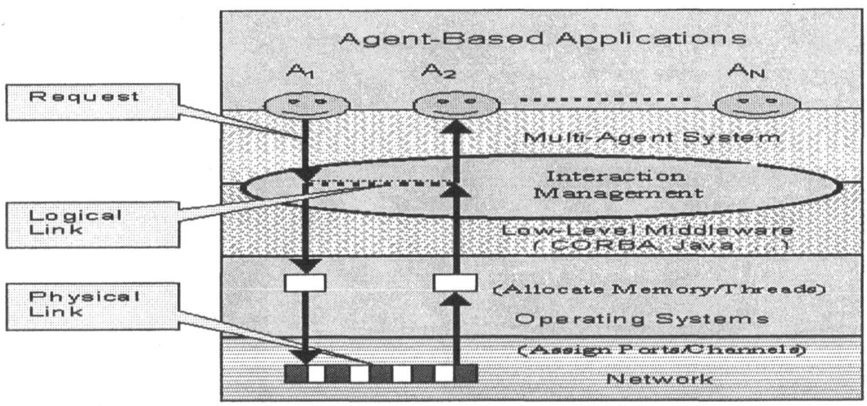

Figure 3. Management of agent-based components.

While Internet technology has developed and used popularly agent technology is still in the development stage.

2.3 Internet-Agent Integration

Agent and Internet technologies can complete each other and are a perfect integration for building autonomous, complex, and distributed business and engineering systems.

An Internet agent-based system (IAS) consists of one or several server sites $\{S_1,S_2,..S_K\}$ supported by Web servers and a dynamic number D of client sites $\{C_1,C_2,...C_D\}$ having Web browser as shown in Figure 4. Each of these sites runs a middleware which supports a multi-agent system (MAS). If these MASs are not the same they must be compatible. In IAS, a user can specify the task with his preferences or requirements via a user-friendly web interface, for instance to buy air tickets, reserve hotel rooms, theater tickets, ... for a vacation. These information will be accepted and analyzed by an agent A1 at a client site or at a server site. Based on the user requests agent A1 may contact seller agents A2, A3, A4 at different sites and may also visit web pages providing air tickets, hotel rooms, and theater tickets separately. By exchanging data with these agents and extracting data from the Web pages A1 know where the offers are, A1 can compare and choose the best combinations of these offers satisfying user critical requirements about time, price, and locations.

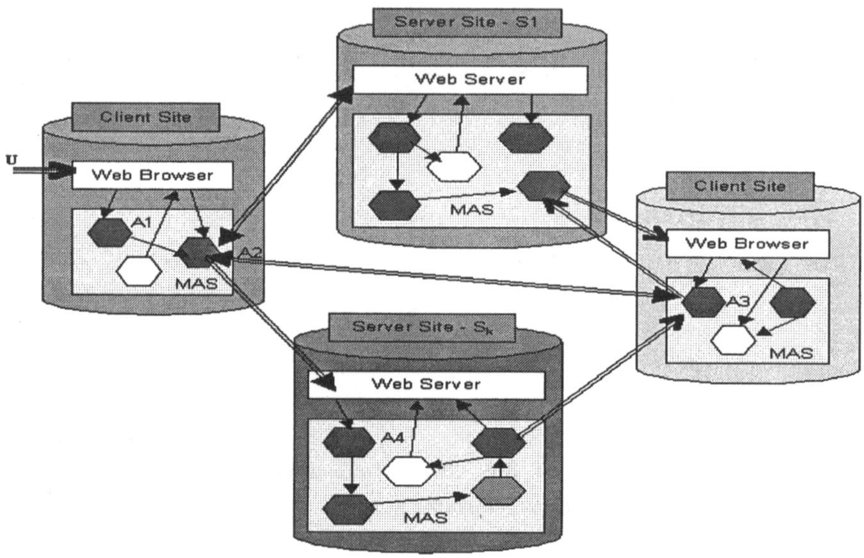

Figure 4. An Internet agent-based system (IAS).

Thus, instead of spending a huge amount of time serfing the Internet, the users in IAS can just specify the tasks to the agents and all the rest will be done automatically. More examples of IAS are described in Section 5.

Notice that some agents are created by users and some by the IAS. Some agents work locally while some can interact with other agents at remote sites. While users have access to the web-interface-agents they do not have direct access to many other agents even though these agents work for them. In IAS the team of agents which implements a user request is formed dynamically and may reside at a number of different locations.

3 Classifications of Internet Agent-Based Systems

We will analyze Internet agent-based systems from several different points of view and provide several classifications of IAS by different criteria. Internet agent-based systems can be classified by their purposes, the task types, structures and architectures, or by agent types.

3.1 Task Paradigm

Building a general-purpose IAS, which can solve any problem, is extremely difficult and there is no such system as yet. IASs are usually specialized and they can be very different depending on the purpose. IASs can be divided into the following categories:

- Information IAS (I-IAS)
- Monitoring/Controlling IAS (MC-IAS)
- Management IAS (M-IAS)
- Computing IAS (C-IAS)

In I-IASs (Padmanabh 1999) tasks are information-driven and include searching, comparing, or combining data in a distributed, heterogeneous, large area. This may be within a LAN, a WAN, or the Internet. The I-IASs are more powerful and more effective than regular search engines because the data may be in different formats at various levels and locations. In I-IAS we can build different agents for different formats and locations. A way to simplify agents in processing data on the Internet is to build semantic Web pages.

The primary purpose of MC-IASs (Nardi 1998) is to monitor dynamic processes. For example to display current states, workloads, numbers of jobs in and number of jobs out in each unit manufacturing machine at a highly automatic plan. In order to control dynamic processes MC-IAS could also include agents which are able to make necessary changes of system parameters and can meet requirements of real-time systems.

M-IASs (Pham 2001) provide users with the necessary tools to manage complex systems at an abstract level without knowing or handling the details. These systems are more autonomous and more sophisticated than

MC-IAS. An example of M-IAS may be an investment system which can help users to trade stocks or make investment decisions based on information available on the Internet. This system can process and analyze data and then generates automatically the best trading variants or investment options to meet the users goals without having users to examine all the details given in portfolios and reports of the companies.

A C-IASs (Yamamoto 2000) provides the user with a virtual and powerful machine for computation. The C-IAS represents and combines heterogeneous resources distributed at different locations. It hides all the network complexity from the user and makes resource locations and names transparent. C-IASs can be used for scientific calculations or for computer network administration.

3.2 Agent Type Paradigm

Mobility is the ability of agents to move unaided from one site to another. Agent mobility can have significant effects on the architectures and the performance of IAS. In term of mobility IAS can be divided into the following categories as shown in Figure 5.

- Mobile IAS
- None-Mobile IAS
- Semi-Mobile IAS

In Completely-Mobile systems (Lange 1999, Suri 2000, Wong1999), agents are free to dispatch to other hosts and retain their own ID. Each host has its own manager. This manager is completely independent and does not share its data with managers of other hosts (Figure 5.a). An agent can interact directly only with agents located at the same host.

In Semi-Mobile systems (Nissen 1999) agents move to another host by creating child agents by remote creation. Thus when an agent migrates to another host its own ID changes. Managers of all hosts share the same agent directory (Figure 5.b). An agent can interact directly with the other agents if these agents are at the same host or through its children which are at the same host.

In None-Mobile systems (Schmidt 1998), even though agents can be created remotely they are unable to move to another host. The management is carried out in a centralized way at higher level. One of hosts may be devoted to be a management server for the system. An agent communicates with the others in the same way regardless of locations. The general communication structure may be illustrated as shown in Figure 5.c.

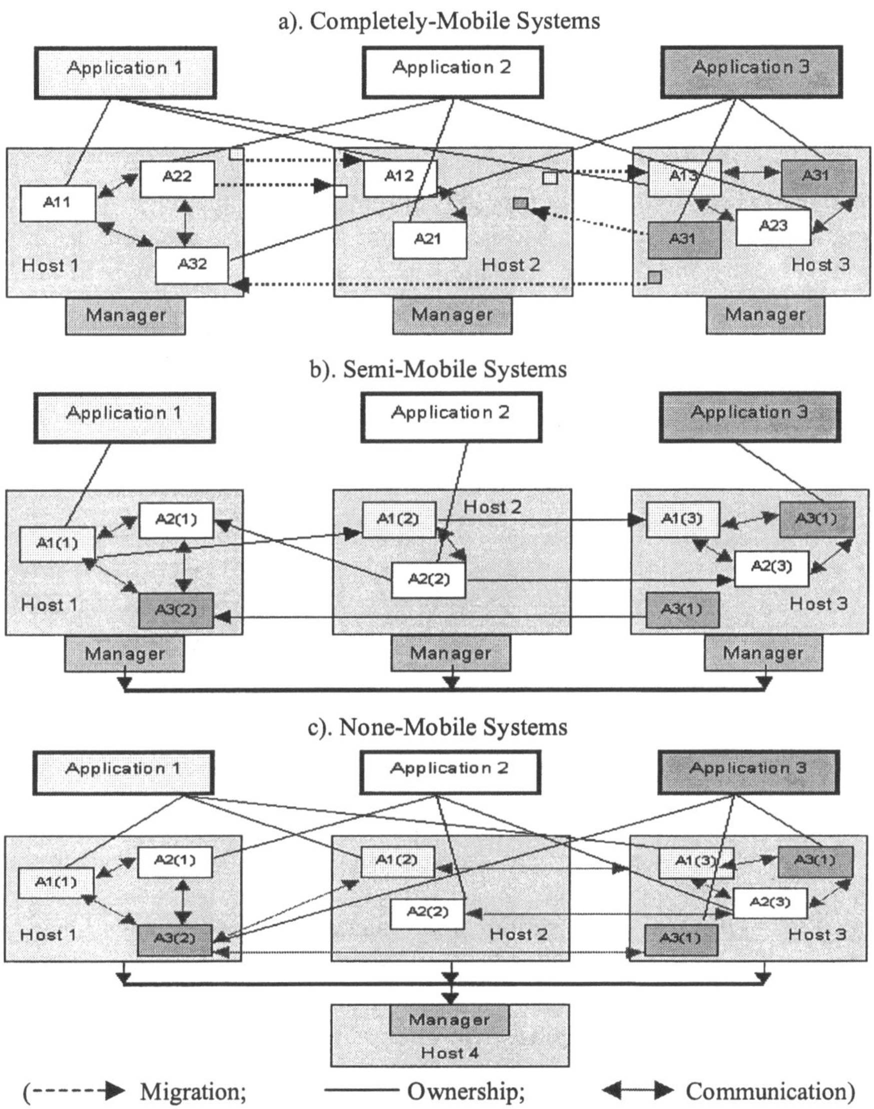

Figure 5. A Classification of IAS by Agent Mobility.

3.3 Architecture Paradigm

The various architectures of Internet agent-based systems can be divided into the following categories: *Yellow-pages* (**YP**) (Mammen 1997), *Contract-Net* (**CN**) (Smith 1980, Soto 2000), *Pattern-based* (**PB**) (Kiniry 1997), and *Point-to-point* (**PP**) (Cassandra 2000). These can be illustrated respectively as shown in Figures 6, 7, 8, 9.

In Contract-Net: In this architecture cooperation is specialized into negotiation and auction. There is a so-called auction manager and the agents are the bidders. All agents declare and send their offers to the manager using Token-Ring topology (Figure 6). The manager maps two agents for a transaction. Messages from the agents are sent to and controlled by the auction managers only and are not known to each other. The agents are free from matching process which is done by the auction manager. However the use of token-ring topology may slows down the system as manager can "talk" only with a limited number of agents at a time and therefore may decrease the speed and quality of matching, since it can not catch all dynamic bids at a time.

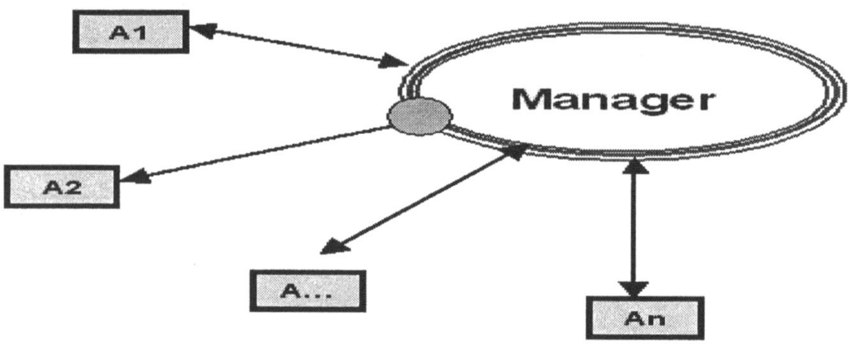

Figure 6. IAS with Contract-Net Architecture.

In Yellow-pages: Cooperation and interaction are arranged by a Manager and all agents share a database called yellow pages. Agents advertise their services and a unit called Matchmaker then maps agents, according to their advertising. The interaction network then consists of pairs between the mapped agents (Figure 7). Compared with CN, this method is more flexible and more interactive as an agent can access and change yellow pages at any time, do not have to wait to his turn as in

CN. However the overheads increase as the manager needs to retrieve and match the data in the yellow pages frequently since the data may be changed by agents at any time.

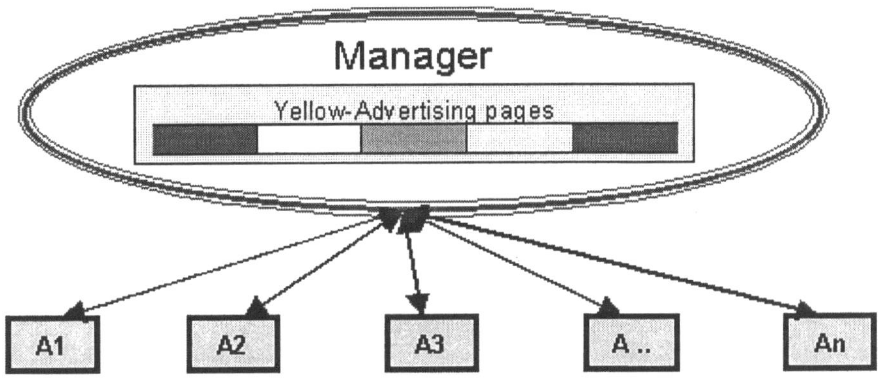

Figure 7. IAS with Yellow-Pages Architecture.

In Pattern-based: All agents are divided into groups and each agent communicates with a group of agents rather than with a particular target agent which may not be known. The network consists of one-to-many connections (Figure 8). Each group of agents represents a service category and has a local manager. All the agents within the group can post/read messages to/from a common "blackboard". As each agent communicates with a group, the overall workload of the system is less than point-to-point if the number of services is much less than the number of agents. The workload for each agent is also reduced as the matching areas are less as it concerns only agents who have the same interest. However there is no privacy as messages are available to all members of the given group.

In Peer-to-Peer: Each agent can exchange information with all other agents at any time, based on their need. The management is the simplest and most flexible as each agent is in charge and is able to create an interaction channel to the others (Figure 9). The overall interaction network is however complex and can overload the host. Besides, each agent must be powerful and have life time long enough so that it can "meet" all the others and find those match. As the matching is done by both sides the system may be redundant.

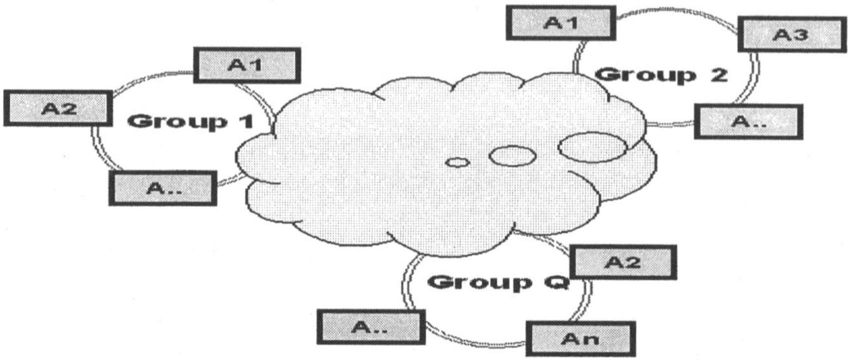

Figure 8. IAS with Pattern-Based Architecture.

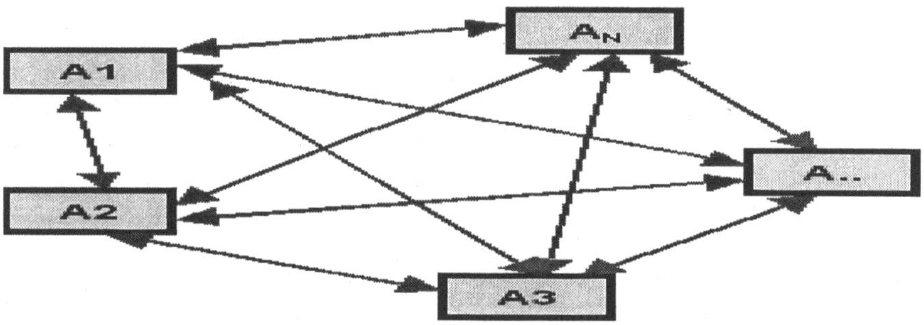

Figure 9. IAS with Point-to-Point Architecture.

Thus, the existing agent-based systems are able to provide flexible interaction in miscellaneous contexts and different formats: one-to-group, group-to-group, one-to-one, group-to-one. In building a cooperation relationship, matching is an important issue. In matching agents should be able to recognize their partners among many other agents. In PP, matching is left to each of agents. In PB, the matching areas are reduced to within a group which decreases the workload of the agents but limits their choice. In CN and YP, opposite to PP, the manager carries out matching alone and therefore agents can not customize the matching criteria or matching degrees.

Among the examined models only CN and TabiCan (Yamamoto 1999) control the number of interactions at a time. The other models create interaction channels using standard interprocess mechanisms and schedulers of low-level middlewares or operating systems. These do not consider the limited capacity of hosts and therefore would cause systems to crash when the number of interaction is very high and overloads the capacity of the system, as reported in (Yamamoto 1999, Pham 2000). We also found that none of these methods consider agent time constraints in organizing agent interaction. Therefore, when the number of agents is high agents in such systems unfairly suffer convoy effect and resultant starvation (Rana 2000).

4 Communication and Coordination in Internet Agent-Based Systems

In Internet Agent-based systems communication is the main way for software agents to interact with each other and coordination of IAS components is carried out via agent interactions. In building an infrastructure for agent communication and component coordination of an IAS we may be involved with the following issues: (i) communication languages that can be understood and used by all agents; (ii) procedures for each agent to send and receive messages or to start a communication; (iii) planning and scheduling systems of each agent which defines when and what data should be exchanged. (iv) system architectures with mechanisms and protocols which create communication links and deliver messages.

We can develop a specialized (Reeves 1999) agent communication language for each IAS using AI techniques or we can use a standard language such as KQML (Yannis 1999). While the design of a specialized communication language for a particular IAS may be a time consuming process, it is always more efficient than using a standard language. If the particular IAS is for a specific problem only then using a standard universal language with its complex grammar and vocabulary can make the agents overloaded with parsing and analyzing messages.

Mechanisms for sending or receiving messages are built as parts of agent actuators. In IASs these procedures are supported by the Internet engine of the agents (Figure 2) and are built using techniques of distributed computing. We will describe them in detail in Section 5.

The planner or scheduler is a part of the agent thinking engine and is responsible for making decision about the order and the contents of messages to be sent or for reading the received messages. These agent components are built using AI techniques as described in Section 5.

The system communication architecture of an IAS plays the role of a broker to support agents exchange messages. In building a system communication architecture for IAS scalability, performance, and reliability are critical issues (Shehory and Sycara 2000, Pham and Tran 2000). We will therefore describe the functions, the problem, and the requirements in building communication architecture for Internet agent-based systems. Assume that agents communicate with each other by exchanging messages of various sizes. The communication management system of an IAS has the following functions:

1. Accept and store messages from agents.
2. Read agent requests by extracting the headers of messages to define the type and the destination.
3. Create a communication link for each request or message. The link could be a queue, a mailbox, or a communication channel depending on the agent requests.
4. Hold the communication links until the messages are accepted by the destination agents or until the delivering time expired.

Notice that each logical communication link created in MAS level corresponds to a physical link (Figure 3). Each communication link may require memory, threads, and ports, etc. which are allocated by the operating systems and the network systems.

The problem of designing a communication management model for IAS can be described as follows. Assume that at a moment of time t the total number of agents in the system is N. The particular N agents { A_1, A_2, ..., A_N } are distributed in the system between the H hosts. The number of agents in a host H_j is:

$$D_a(H_j) = \alpha_j \text{ , where } j = 1..X$$

A map function μ will define where an agent A_i is:

$$\mu(A_i) = H^k \text{, where } i = 1..N \text{ and } H^k \in H$$

Each agent A_i may generate a number δ_i of messages, i=1..N. Thus, the total number of messages or communication links in the systems is:

$$T_{Tm} = \sum_{i=1}^{N} \delta_i$$

These messages are distributed within X hosts as follows:

$$T_m(H_k) = \sum_{i=1}^{N} W_i^k \times \delta_i \quad \left| \begin{array}{l} W_i^k = 1 \; if \; \mu(A_i) = H_k \\ W_i^k = 0 \; if \; \mu(A_i) \neq H_k \end{array} \right.$$

Given a computer system H with a fixed capability. It consists of X hosts or servers: $H=\{H_1, H_2, .. H_X\}$. Each host can handle a fixed maximal number $MaxL_{Hj}$ of queues, communication channels, or mailboxes:

$$MaxL_{H_j} = \Psi(H_j), \quad where \quad j = 1..X.$$

The network can be represented by a graph $G_{net}=\{H,\Psi,C,W\}$ as in Figure 2, where $H=\{H_1, H_2, .. H_X\}$ is the set of nodes representing X hosts, $\Psi=\{\Psi(H_1), .., \Psi(H_X)\}$, where $\Psi(H_i)=MaxL_{Hj}$; $C=\{(C_{k,l})| \; k=1..X, l=1..X, k\neq l\}$ is the set of connections between hosts H_k and H_l, and $W=\{(W_{k,l})| \; k=1..X, l=1..X, k\neq l\}$ is a set of transmission capacity of the corresponding connections in C.

Assume that the communication of agents in MAS is managed by a model MO. This model MO could be considered as a virtual device implementing a meta-function COM(A,H) which includes a function $\Phi c(k,l)$, where k=1..X, and l=1..X. This function $\Phi c(k,l)$ returns a value which is the number of communication links that are created by the given communication model MO in the connection $C_{k,l}$ between host H_k and host H_l. In the case when k = l, $\Phi c(k,l)$ returns the number of communication links that are created in the host H_k by the given communication model MO.

Definition-1: Network congestion

Assume that $f_C(k,l,t)$ is the network congestion function of the connection between host H_k and host H_l at time moment t. When $k\neq l$, ***if*** the number returned by function $\Phi c(k,l,t)$ where k=1..X and l=1..X, i.e. the number of created communication links, is more than $W_{k,l} \in W$

which represents the transmission capability of the connection between hosts H_k and H_l , ***then*** there is a network congestion in this connection between H_k and H_l and we have $f_C(i,t)=1$.

The total number of network congestion T_{cong} can be defined as:

$$T_{cong} = \sum_{k=1}^{X} \sum_{l=1}^{X} f_C(k,l,t) \left| \begin{array}{l} f_C(k,l,t) = 1 \ \ if \ \ \Phi c(k,l,t) > W_{k,l} \\ f_C(k,l,t) = 0 \ \ if \ \ \Phi c(k,l,t) \le W_{k,l} \end{array} \right.$$

<u>*Definition-2:*</u> Host failure

Assume that $f_H(i,t)$ is the host failure function of the host Hi at time moment t. When k=l, ***if*** the number returned by function $\Phi c(k,l)$ where k=1..X and l=1..X, i.e. the number of created communication links, is more than $MaxL_{Hk}$ which represents the maximal allowed number of the communication links in host H_k, ***then*** there is a failure at the host H_k and we have $f_H(i,t)=1$.

The total number of host failures T_{fail} can be defined as:

$$T_{fail} = \sum_{i=1}^{X} f_H(i,t) \begin{cases} f_H(i,t) = 1 \ |\Phi c(i,i,t) > \Psi(H_i) \\ f_H(i,t) = 0 \ |\Phi c(i,i,t) \le \Psi(H_i) \end{cases}$$

Notice that, in both cases, the system is overloaded and may fail as its capability can not hold the activated agent communication links. However, it may not be the agent demands but the decisions of the communication management system, exp. when and by which routes messages should be transferred, which generates the system failures.

The system reliability depends on the total number of host failures T_{fail} and the total number of network congestions T_{cong}. It can be measured as the following:

$$\text{Re}(H, A, t) = \frac{T_{res} - T_{fail} - T_{cong}}{T_{res}}$$

Where, T_{res} is the total number of hosts and activated connections which can be defined as:

$$T_{res} = X + \sum_{k=1}^{X} \sum_{l=1}^{X} \Omega(k,l) \left| \begin{array}{l} \Omega(k,l) = 1 \ | \ W_{k,l} > 0 \\ \Omega(k,l) = 0 \ | \ W_{k,l} = 0 \end{array} \right.$$

As stated in (Rana and Stout 2000), scalability of MAS could be considered in different dimensions: numbers of agents; complexity and volume of agent activities; and agent diversity. In this chapter we focus on the scalability of MAS in terms of: (i) numbers of agents in the global system across multiple platforms; and (ii) total number of messages.

Then, the requirement and goal in managing communication for IAS in term of scalability is to design a model M* so that:

$$\int_{Ao}^{A^*}\int_{0}^{\infty} \mathrm{Re}(H, A, t)\, dt\, dA \rightarrow \min$$

Where, A* is the possible largest IAS (among the other parameters of A*, N* is the possible maximal number of agents, which could be 1,000,000 or even more in massive IAS) and Ao is the possible smallest IAS; dt represents time change; and dA represents any change of the multi-agent system A which could be a change of number of agents or a change of their configuration when agents migrate, etc.

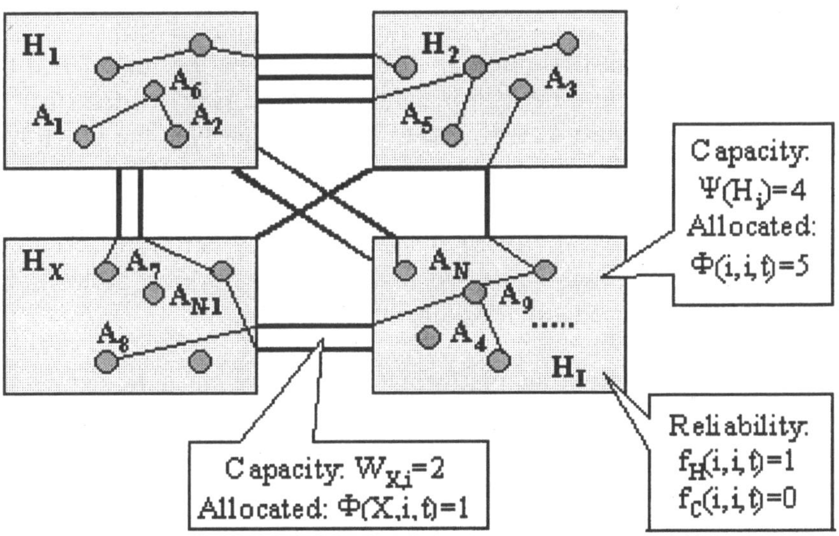

Figure 10. Resources Allocation for Communication in IAS.

Examples of IAS communication systems are provided in Section 6.

5 Procedures and Technologies for Designing Internet Agent-Based Systems

In order to develop an IAS we may need to employ AI techniques and technologies of distributed computing and can follow the following steps:

1. Represent the Problem and its Solution
2. Build IAS Architecture
3. Define Computing Platform
4. Design Agents
5. Build Communication System
6. Coordinate Components
7. Analyze and Evaluate IAS

(1) The first step in designing an IAS is to state the problem for which IAS need to solve. Then its solution or algorithm should be *modularized* and represented with a maximized *parallelism*. We can use graphs or the Unified Modeling Language (UML) and other *software engineering techniques* to build and represent the solution.

(2) The next step in designing an IAS is to define the model and the architecture of the Internet components and the agents as described in Section 2.3 and Figure 4. We need to define: What type the given IAS is, as described in Section 3.1; How many server sites $\{S_1,S_2,..S_K\}$ are needed; What the scaling degree is, i.e., what the maximal number of client sites $\{C_1,C_2,...C_{Dk}\}$for each server site is; What the functions of each site are. In other words we need to define which parts of the solution will be carried out at which site.

(3) In defining computing platform for IAS, we need to define the middlewares which support the Internet components and the software agents. As analyzed in Section 2.2 we need to define which Web servers and other Web supporting middlewares should be used at each server sites and which Web browser should be expected or developed. For software agents we can (i) use an existing multi-agent systems such as IBM Aglets, Mitshubishi Concordia, ... or (ii) develop a specialized multi-agent system for this particular IAS only.

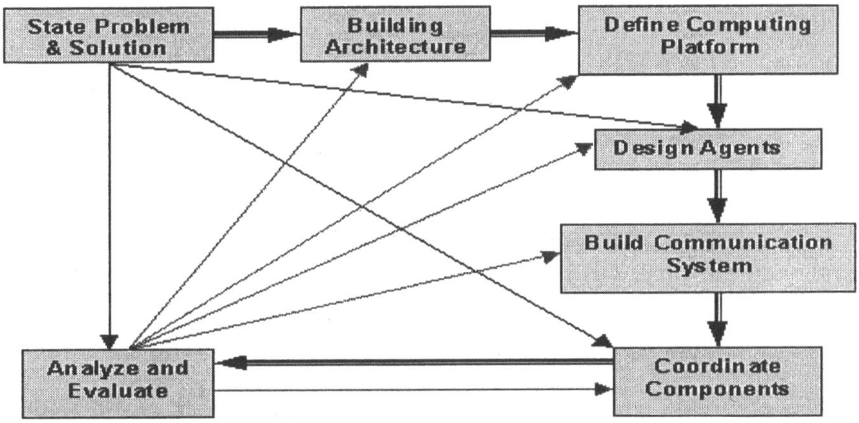

Figure 11. Design Steps for IAS.

(4) Regardless of which approach has been chosen in the previous steps we need to define (i) how many agents are needed in the given IAS; (ii) functions of each agent; (iii) how these agents should be distributed in the system, i.e. at which sites; (iv) the structure and components of each agent and its cycle, as described in Section 2.2; and (v) the agent types as described in Section 3.2. Agent behaviors including its actions state transitions can be represented using finite automata or Petri nets.

In designing the thinking engines of the agents the following AI fields and techniques may be used or involved:

- theoretical planning,
- decision support systems,
- fuzzy logic,
- genetic algorithms,
- neural networks,
- natural language processing

In developing the Internet engines the following technologies of distributed computing can be used:

For Non-Mobile-Internet agents: (for Data Transmission)

- Sockets
- Java Servlets
- CGI, environment variables
- http tunneling

For Mobile Internet agents: (for Dispatching or Migration)

- RPC/RMI
- CORBA/DCOM
- D-OS (Stop-and-copy, Dirty Copy, Copy on Reference)

(5) The mains steps in building a communication management for an IAS, as stated in Section 4, are: (i) define a communication language; (ii) message format; (ii) mechanisms for each agent to send and receive messages; (iii) develop a system communication architecture and protocols for managing communication links and delivering messages.

(6) The next step is to coordinate the components of IAS. After having built Internet modules, multi-agent middleware and agents we need to synchronize the modules and the agents so that their coordination is in accord to the given algorithm. A way to simplify this process is to divide the algorithm into transactions. Protecting shared data and keeping multiple copies of the same data consistent are also critical in coordination and can be done using distributing techniques such as semaphores, monitors, or locks.

(7) We need to analyze the designed IAS and estimate it in terms of correctness, performance, reliability, scalability, and security. The analysis can be carried out using probabilistic or modeling-programming approaches. It may lead to many corrections or modifications in the IAS design. This step allows us to avoid errors and improve the design before the components of IAS are implemented.

Only after obtaining a sound design we will implement the IAS and then test it with different scenarios.

6 Internet Agents for System Design

Design and building a complex system may involve many people and require a tremendous amount of data which can be located at different locations. Internet agent-based system may support this process in an efficient and flexible way. The Internet provides physical means to transfer and share data while software agents can gather necessary information, prepare it, and coordinate the designing specialists automatically.

Therefore, there appears ideas about building a meta-IAS which can support the designing of other IASs (Jamali 1999, Shu 1999, Nwana 1998, Parunak 1998). The meta-IAS can generate a frame work, by the design steps described in the previous section, for the designers and then coordinates and combines their work. The IASs for designing should be built based on particular problems for which the target IASs are designed for. For instances, an IAS for designing an E-accounting system or an IAS for designing an e-news service.

In general case, the structure of a designing-IAS (D-IAS) can be illustrated as shown in Figure 12 which has one server site and could have a dynamic number of client sites used by designers at different locations. A general D-IAS implements the following functions:

1. To represent the problem in a user-friendly and dynamic form: there could be a team of agents (PTA) which collects data or parameters from the users about the given problem and then constructs these data, forms the problem statements, and displays it to the designers.

2. To construct the solution for the given problem: there could be another team of agents (STA) which support the designers to build the solution. The complete solution should be divided by task-modules and should be represented using a standard representation language or diagrams and is available at a Web page at the server site so that the designers can examine the whole algorithm and make bids to the modules based on their expertise.

3. To accept the bids from the designers, compare them and then assign and schedule the modules to the designers based on the obtained bids. We may have a team of agents (BTA) which work as an auction broker where the designers are the bidders and the modules are the items to bid. BTA should also work as a scheduler which generates deadlines for modules based on the time constraint and the designers availability. We can employ scheduling mechanisms for distributed and heterogeneous objects such as the ones in (Pham 1997, Pham 1999).

4. To provide the data and necessary information to each designer based on the task-modules which are scheduled to him. This can be done by a team of agents (DTA).

5. To accept the designs for task-modules from the designers and combine them into a single design solution and display it at a Web page at the server site. A team of agents CTA should be responsible for this.

6. To coordinate the designers and provide communication facilities so that the designers can cooperate with each other in their designs.

7. To provide tools and environments for designers to analyze and test their modules separately or the whole design solution. A team of agents ETA should be responsible for this.

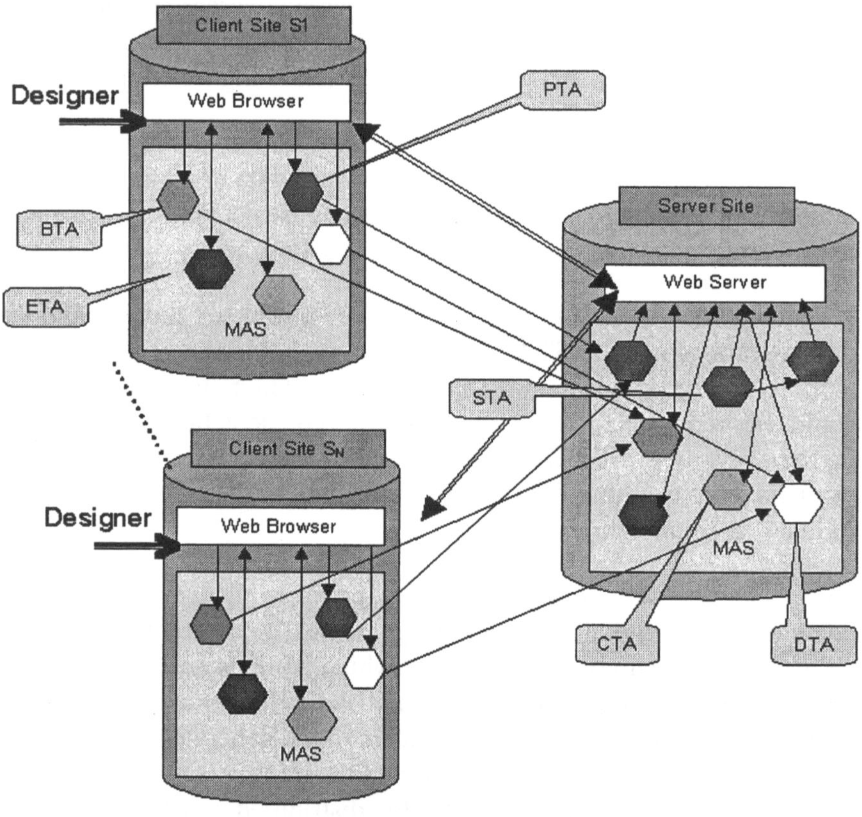

Figure 12. IAS to Support System Designers.

7 Case Studies

Internet agent-based systems as an integration of AI and distributed computing can be applied to many different areas such as e-business, manufacturing management, engineering monitoring and controls. In this section we will describe: (i) an IAS for e-controlling and e-managing inventory in a large and distributed manufacturing system; and (ii) an IAS for a scalable e-marketplace.

7.1 An IAS for Inventory Control

Modern corporations consist of decentralized business units and require multistage production planning. We propose an E-management approach to inventory control and particularly to lot-sizing problem in distributed and multistage production. The proposed solution is an Internet agent-based system where software agents cooperate with each other to implement a genetic gender-based algorithm in order to avoid local optimums in a large and multistage search space. The use of agent technology helps us to reduce computation complexity, speed up planning time, and especially to support distributed business units so that they can coordinate and interact more efficiently and respond more dynamically to changes in E-commerce environment.

7.1.1 Inventory Management Problem and its Solution

A multistage production could be described as the following. A company produces Y products and has X business units which includes: P procurement units; M manufacturing units; and S selling units, X=P+M+S. Procurement units are responsible for buying materials and distribute them to manufacturing units. Selling units are responsible for collecting final products and sell them. A manufacturing unit may use raw materials or middle products produced by other units. Outputs of a manufacturing unit could be inputs for the other units. Each unit can participate in producing several products. All units which involve in producing a product form a product line. We may have W product lines: $\{PL_1, PL_2, .., PL_W\}$. In a product line, the units and their connections are defined by its product schemata $SCHEM(PL_i)$, $i=1..W$. Hence, the production structure of a company is represented by:

- The set of business units or production structure:

- Procurement: $\{PU_1, PU_2, .. , PU_P\}$;
- Manufacturing: $\{MU1, MU2, .. , MU_M\}$; (1)
- Selling: $\{SU_1, SU_2, .. , SU_S\}$.

- The product line configuration: $SCHEM(PL_1), SCHEM(PL_2), .. , SCHEM(PL_W)$.

Assume that for each unit Ui, i=1..X, which participates in W_i product lines, we have a set of inputs $INP_{i,k}$ which we will call local materials, a set of outputs $OUT_{i,k}$ which we will call local products, and a set of production functions F_{ik}, k=1..W_i, which defines the amount of products $PRO_{i,k}$ needed to be produced by the given unit. The unit U_i virtually has its own e-warehouse with V_i items including local materials and local products of all W_i product lines. The costs for holding items are defined by functions $G_{i,k}(v_{i,k})$, where vi,k is the subtotal items for product PL_k, k=1..Wi. The costs for transferring local products from Ui to the others are defined by $H_{i,k}(R_{i,k},U_j)$, where $R_{i,k}=OUT_{i,k}$, k=1..Wi, and Uj is the destination unit. Assume that the production process is carried out by a series of time moments: {1,2, ...,T}. Then, in order to satisfy the given production plan we have:

$$V_i(t) = V_i(t-1) + \sum_{k=1}^{W_i} INP_i^{\,k}(t) - \sum_{k=1}^{W_i} OUT_i^{\,k}(t) \qquad (2)$$

The lot-sizing problem (Pfeiffer 1999, Ben-Daya 1999) in this case is to define a schedule of holding and transferring local materials and products for each of X units. For each time moment t, where t=1..T, a unit should receive a planed number $INP_{i,k}(t)$ of materials, and produce a planned number $PRO_{i,k}(t)$ of products, and output a planned number of its products $OUT_{i,k}(t)$ so that the total cost of storing $V_{i,k}(t)$ and transferring $R_{i,k}(t)$ was minimized. In other words, the given problem can be stated as follows:

Given:

- X business units: $\{PU_1, PU_2, \ldots, PU_P\}$, $\{MU_1, MU_2, \ldots, MU_M\}$, $\{SU_1, SU_2, \ldots, SU_S\}$.
- Production rules described in (2) and production configuration for W product lines given in: $SCHEM(PL_1), SCHEM(PL_2), .., SCHEM(PL_W)$

(3)

- Estimated market demands, i.e. the amounts of products that can be sold at S selling units: $OS_1(t), OS_2(t), \ldots, OS_S(t)$, where t=1..T.

Requirements:
Define a schedule for holding and transferring products, i.e. define the non-negative elements of matrix R[1..X][1..X][1..T] for transferring and the non-negative elements of matrix V[1..X][1..T] for holding for each time moment t, t=1..T, so that:

$$\begin{cases} (*1)\ INP_{i,k} \geq 0, \text{and } OUT_{i,k} \geq 0, i = 1..X, k = 1..W_i \text{ , given } (1),(2),(3) \\ (*2) \sum_{l=1}^{W} \left(\sum_{1}^{T} \sum_{i=1}^{X} V[i][t] + \sum_{t=1}^{T} \sum_{j=1}^{X} \sum_{i=1}^{X} R[i][j][t] \right) \rightarrow MIN \end{cases} \quad (4)$$

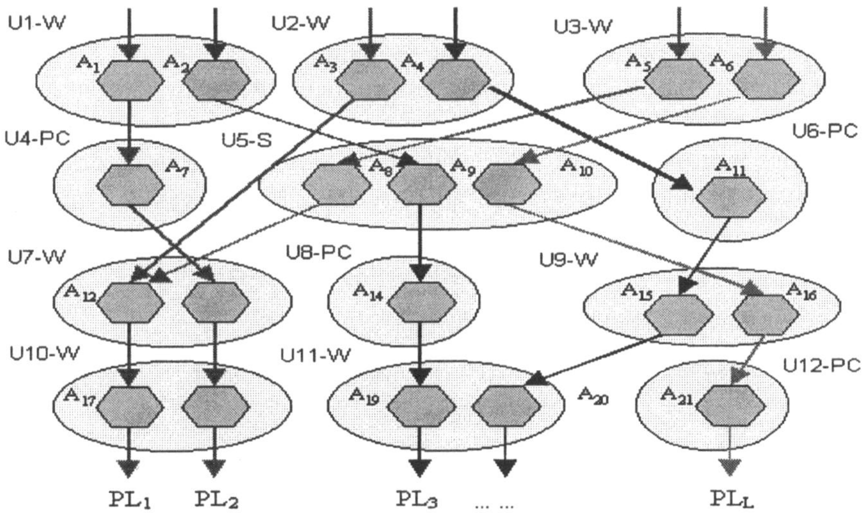

Figure 13. Agent-Based Model for Inventory Management.
W = Workstations; PC = Personal Computer; S = Server machine.

7.1.2 Build IAS Architecture

We propose to use agents in building the production management system in the given problem to reduce its complexity. In our agent-based model (Figure 13), each agent represents a business unit U_i, i=1..X, in one product line PL_k, $k=1..W_i$. Thus, if the given unit U_i participates in W_i product lines it would have W_i agents. Then, in total we would need N agents to support X units:

$$N = \sum_{i=1}^{X} W_i$$

Assume that the given production requirements are defined by (1), (2), and (3). All of data concerning inputs, output, warehouse, and production of the given unit are stored locally and are managed by local agents, one per product, instead of being stored and managed by a centralized center of the company.

The architecture of an agent with its data flow is zoomed-in in Figure 14. Within an agent A_i, i=1..N, at a time moment t, there is $C_i(t)$ amount of local materials comes in and there is $E_i(t)$ amount of local product goes out. Each agent has two main components: e-warehouse and e-manufacture. The e-warehouse stores $IA_i(t)$ amount of intermediate and $OA_i(t)$ amount of a product. The e-manufacture component needs $QA_i(t)$ amount of local materials to produce $PA_i(t)$ amount of local product items. Thus, the initial lot-sizing problem for the whole company is broken down into N identical and much less complicated sub-problems for single agent as follows.

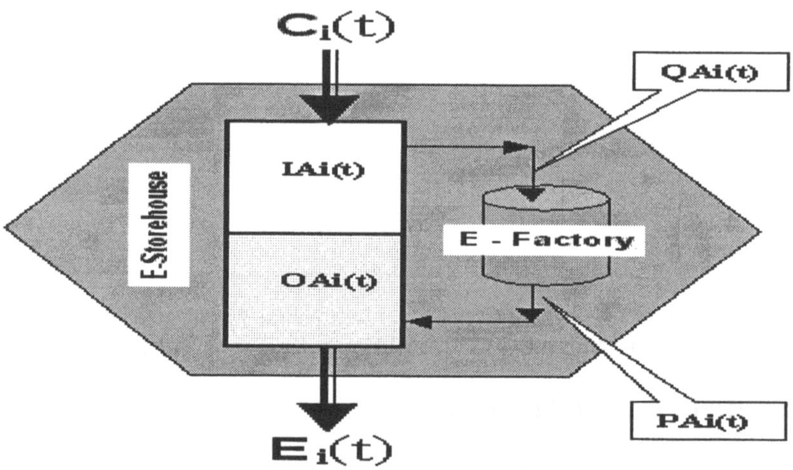

Figure 14. Agent Configuration and its Data Flow.

Given production rules which are described in (1) and the configuration for W product lines: $SCHEM(PL_1)$, $SCHEM(PL_2)$, .., $SCHEM(PL_W)$, we can define the configuration of the agent-based model for production management as in Figure 13. Then, knowing the market demands: $OS_1(t)$, $OS_2(t)$,..., $OS_S(t)$, where t=1..T, and the agent configuration (Figure 2) we can backtrack $PA_i(t)$ for each agent A_i, i=1..N, at each time moment t=1..T. From production functions F_X^k, k=1..W_X for each unit U_X, x=1..X, we can easily infer the production ratio FA_i for each agent A_i, i.e:

$$PA_i(t)=FA_i(QA_i(t)) \text{ or } QA_i(t)=FA^*_i(PA_i(t)).$$

Next, the conditions for satisfying these production requirements for agents can be described as:

$$\begin{cases} IA_i(t) = IA_i(t-1) - QA_i(t) + C_i(t) \geq 0 \\ \qquad \forall t = 1..T, i = 1..N \\ OA_i(t) = OA_i(t-1) + PA_i(t) - E_i(t) \geq 0 \\ \qquad \forall t = 1..T, i = 1..N \end{cases} \tag{5}$$

The requirement for agent lot-sizing is to define $E_i(t)$, i=1..N, t=1..T so that:

$$\sum_{t=1}^{T} \sum_{i=1}^{N} (VA_i(IA_i(t) + OA_i(t)) + RA_i(E_i(t))) \tag{6}$$

was minimum.

Notice that, from $E_i(t)$, i=1..N, t=1..T and using agent configuration (Figure 14) we can define $C_i(t)$, i=1..N, t=1..T since output of an agent forms input of other agent. Then, as $PA_i(t)$ and $QA_i(t)$ are also known, we can recursively define $IA_i(t)$ and $OA_i(t)$ using the equations in (5). Thus, we can check (5) and (6) recursively by time and in parallel by agents.

7.1.3 Define Computing Platform

The given IAS is a fine-grained distributed system but not a large-scale one. Therefore we may need only one server site which will be one of a node in Figure 1. Besides hosting the agents for this node this server site provides a Web-based interface of the inventory data located at different machines for all the users in the systems. It may also back-up these distributed data at the server site so that if one machine fails the server can recover its data when it restarts or when it is replaced by some other machine.

7.1.4 Design Agents

In the proposed agent-based model, the key point is to find a set EA of $E_i(t)$, i=1..N, t=1..T which satisfies (5) and (6). Recall that $E_i(t)$ is the amount of local products which should be transferred at the time moment t, from a unit U_k, $k \in [1..X]$, which is partly managed by agent A_i, to some other unit U_h, $h \in [1..X]$, which is partly managed by agent A_j. The connection A_i-A_j can be defined using the given product line configuration, for instance the agent connection A_6-A_{12} from unit U_5 to unit U_7 (Figure 13). The matrix EA then represents a plan or a schedule of inventory transfers and can be depicted as in If $E_i(t)=0$ that means there should be no transfer from agent A_i at the time t.

We use genetic algorithm (GA) to define optimal values of elements in EA so that we can avoid local optimums, because the search space, whose size is reduced to $Z^{(T*N)}$, is still very large and has multiple levels. In order to speed up the searching time with GA we build a concept called agent gender and will use it to abstract search levels. In the given case study agent gender represents the transfer characteristic of an agent at a time. Instead of generating or reproducing strings, which represent $E_i(t)$ from EA with Z values for each digit in the strings, as in standard GA, we first mark a gender from {0,1} for each digit and then estimate and reproduce strings based on their digit genders rather than on their values. Anytime when a replenishment is needed for an agent A_j the gender of the previous agent A_i should be "1," if gender of A_i is "0" that means there should not be any transfer at the given time t. Recall that A_i and A_j forms an agent connection in the agent chain of the given product line.

This modification of GA using the gender concept simplifies the string format and reduces the complexity of agent local task from $Z^{(T*N)}$ to $2^{(T*N)}$. Now we can work with the gender map of string EA, which is an abstract image of a string, rather than the string itself. Using this abstract notation we can skip obvious non-optimal generations and therefore reduce planning time. The gender map $G[N,T]=\{\{g_1^1, g_1^2, \ldots, g_1^T\},\{g_2^1, g_2^2,\ldots, g_2^T\},\{g_N^1,.., g_N^{T-1},g_N^T\}\}=\{0,1,0,0, \ldots 0,0,1,1\}$ has N*T binary digits while the full-schedule string $EA=\{E_1(1), E_1(2), .., E_1(T), .., E_N(1), .., E_N(T)\}$ would have $N*T*Log_2(Z)$ binary digits. The full schedule of inventory replenishments EA can be defined using its gender map G. Notice that the gender map $G[N,T]=\{g_1^1, g_1^2, .., g_1^T, g_2^1, .. g_2^T, .., g_N^{T-1}, g_N^T\}$ tells us: (i) *when* (by t, t=1..T) and (ii) *where* (by i, i=1..N) the replenishments are made without specifying what is the amount of a replenishment for each agent. However, this hidden data can be calculated recursively as follows. Starting from time 0, assume that the first "1" in the gender map G is g_K^R. This "1" represents the first transfer of the corresponding agent A_K and it can be defined as the subtotal accumulated amount of local products that the given agent A_K has produced from time 0 to time R. Similarly, we can define the amounts of the first transfers for other agents by the positions of their first "1"s in each page of the gender map G.

After each transfer, products are accumulated again and are passed to the next agent in the product chain. Therefore, we have:

$$OA_i(t) = OA_i(t-1) + PA_i(t) - E_i(t) = 0$$

Thus, for each next transfer of G[N,T]={0,1,0,0, ... 0,0,1,1}, we can recursively find the corresponding transfer amount as the following:

$$E_i(t) = OA_i(t-1) + PA_i(t) \tag{7}$$

Then, we can evaluate the total cost or the efficiency of this full schedule or the given gender map by (6). Thus, GA mutation and crossover operations are performed over the map of agent genders rather than over the string-values of the full schedule.

7.1.5 Build Communication System

We use the existing Peer-to-Peer architecture for agent communication as the number of agents in this IAS is not very large. When an IAS is not a large-scale system Peer-to-Peer architecture is the best solution as it provides shortest response times.

7.1.6 Coordinate Components

In order to coordinate the agents we divide the inventory management into transactions based on product lines. Each product line represents a transaction thread. The agents connect with each other in an assembly line fashion and in accord to the configuration of the product lines. We use software locks to protect the shared data. In order guarantee the consistency of multiple copies of the same data we use the Entry Consistency model (Gali 2000).

7.1.7 Analyze and Evaluate IAS

The complexity of lot-sizing problem for each agent is $Z^{(T*N)}$. The total complexity of this approach is the sum of agent's complexity and the overhead of sending data from one agent to the other in a product line. The overhead depends on the product line configurations, especially their lengths, and on the data transmission speeds via networks. In theory, with an assumption that the data transmission time is zero, it is $MaxL*Z^{(T*N)}$ where MaxL is the maximal length of product lines and may range in [1,N]. However, this theoretical estimation on complexity may not show the real picture because the data transmission time is not zero.

In order to estimate the performance of the proposed agent-based genetic method AGGOA we simulate and compare it with a standard lot-sizing method (SM) which uses step-down allocation mechanism (Pfeiffer 1999) and a standard genetic algorithm SGA without agents. We use two criteria : (i) computation time Tc to evaluate performance and (ii) response time Trp to evaluate quality of service. We measure the computation times of the methods when they achieve the same quality of solution, i.e. approximately the same cost of holding and transferring calculated by (6). The results are shown in Figure 15, the time differences between AGGOA and SGA are recorded while Z, the size of the domain, is varying from 1 to 10 values. The simulation results show us that AGGOA can find an inventory schedule of the same quality as by SGA for a significantly less time. The more Z is the more it shows.

In evaluating the quality of service we measure how quickly a new plan of inventory management is found when there is a change in production configuration or in market demands. The response time Trp is the period of time which is needed for each method to modify the existing inventory management plan since the change occurs. Response times of AGGOA and SM are recorded while the number of time moments T is varying from 1 to 12 slices. The simulation results show us that

AGGOA can modify its plans for less time than SM can. This trend continues when the number of time slice is more than 12.

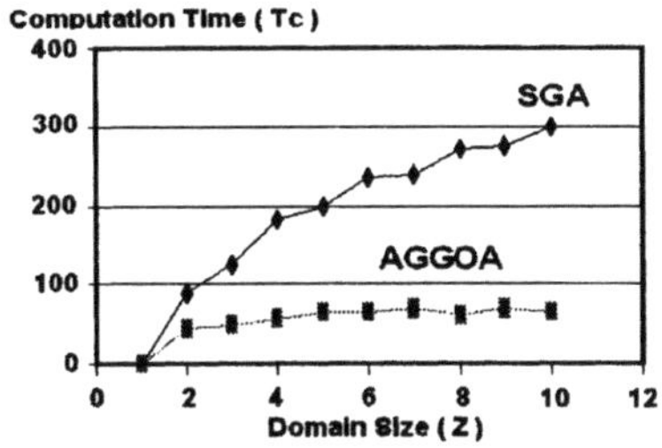

(a)

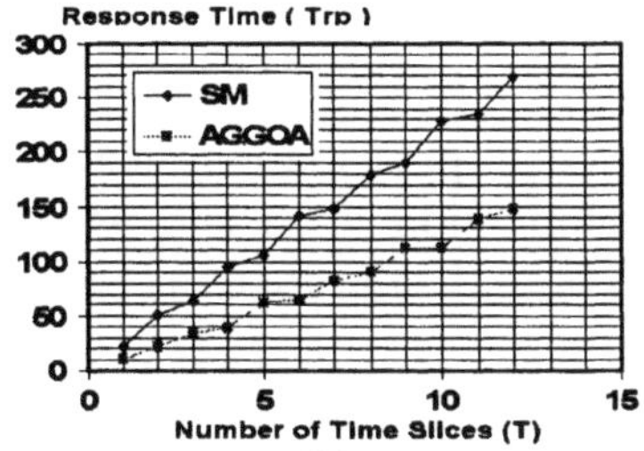

(b)

Figure 15. Simulation results.

The obtained results in Figure 15.a can be explained by the fact that the standard genetic algorithm SGA heavily depends on Z since it represents a transfer plan by a full-schedule string EA = $\{E_1(1), E_1(2), .., E_1(T), .., E_N(1), .., E_N(T)\}$ as usual, using values of the amounts to be transferred. Meanwhile, AGGOA's computation time does not change while Z increases. That is because AGGOA uses gender marks of agents to represent transfer plan and therefore does not depend on Z. We also obtain better results with AGGOA in term of response time (Figure 15.b) because when there is a change, either in production configuration or in market demands, only agents which participate in the given product line modify their data, not all the agents in the system. On the other hand, the proposed method AGGOA will not be better than SGA or SM for the problems with unstructured search space where the gender concept can not be applied. This is because in unstructured search problems where the distribution is random heuristics do not work and a standard GA would be the best solution.

7.2 An IAS for E-Marketplace

We use the IAS concepts to build an e-marketplace with N buyers and sellers who trade M kinds of goods and services. Each buyer or seller owns an agent Ai, i=1..N. The agents should communicate with each other to define the partners which then will negotiate to reach a deal.

7.2.1 E-Marketplace Problem with Special Focuses and Its Solution

Building an E-marketplace (Janin 1999, Senn 1999) involves many different issues such as negotiation, optimization, communication, knowledge presentation, learning, etc. In this chapter we focus on communication issue and will keep the other parts simple. Then the requirements and goals in building an E-marketplace can be stated as follows:

Given: n agents A1, A2, …, An with M categories services S={S1, S2, .. S_M}. Each agent Ai , i=1..n, characterizes by a set of data (Ti, Pi, Di), where:

- Ti is the period of time for which the agent is scheduled to live or agent life time.
- Ri is a set of agent requests, Ri={R_i^1,R_i^2, … R_i^{Ui} }={(S_i^1, tR_i^1),…, (S_i^1, tR_i^{Ui})}, which shows which services from S the given agent Ai is interested in, and deadlines of requests tR_i^k, where k=1..Ui, Ui is the number of requests of agent Ai;
- Di is other data such as sizes of messages, message box's address etc.

Requirements: Design a model and mechanisms for agents {A1, A2, …, An} to exchange messages based on their trading interests given in Ri, i=1..n. The goal is to guarantee reliability while maintaining good performance and quality of service in term of response time.

7.2.2 Build IAS Architecture

In order to manage a large number of trader-agents we propose to build a broker-team of middle agents. Arguments for choosing this approach, instead of building a centralized management is that it helps us to deal with bottleneck problem as the management is *distributed*. Besides, the use of middle agents, with their intelligent and autonomous features, to manage regular agents with the same features also seems to be more *suitable* compared with the use of traditional mechanisms as in non-intelligent systems. With middle agents we can take advantage of the fact that both middle agent and regular agent may use the same language or standards and therefore it is easier and more efficient to proceed agent requests.

The team of middle agents for communication management has three layers and consists of:

- a supervisor-agent (SA);
- a number of teller-agents (TA);
- a number of router-agents (RA).

The team architecture is shown in Figure 16. Requests of regular agents A1, A2, ..., An are redirected by *router-agents* RA to teller-agents TA according to agent interests. A router-agent RA serves a given number of agent requests. Each *teller-agent* TA serves a service category. It divides agent requests of the same category into consumers and providers. Then, the teller-agent creates a channel between any two potential partners and makes them active periodically by the rules of a special mechanism which works as a rotary dining table in Chinese restaurants. This model works similarly to how foods are provided to consumers in restaurants. While router-agents RA_1, RA_2, ... play the roles of waiters, who show customers their seats, teller-agents TA_1, TA_2, ... together play the roles of the rotary dining table at the Chinese restaurants. However, the difference is that the foods are brought by some of the customers (providers). These customers will receive "plates" on the table to put their foods while the others (consumers) will get the "chairs".

The *supervisor-agent* SA constantly takes control over the total number of communication channels, or number of "plates" on the table and number of "chairs," based on the currently available resources. SA also adjust the number of router-agents, based on the number of agent requests, and the number of teller-agents, based on the number of services categories. SA, TA, and RA are described in more details in next sections.

7.2.3 Define Computing Platform

As the number of trader-customer-agents may be very high we should use as many server sites as possible. Besides hosting the customer-agents each server site runs router-agents or teller-agents and also provides a Web-based interface for all the users in the systems. One or several server sites will host the supervisor-agent and its identical copies for back-up.

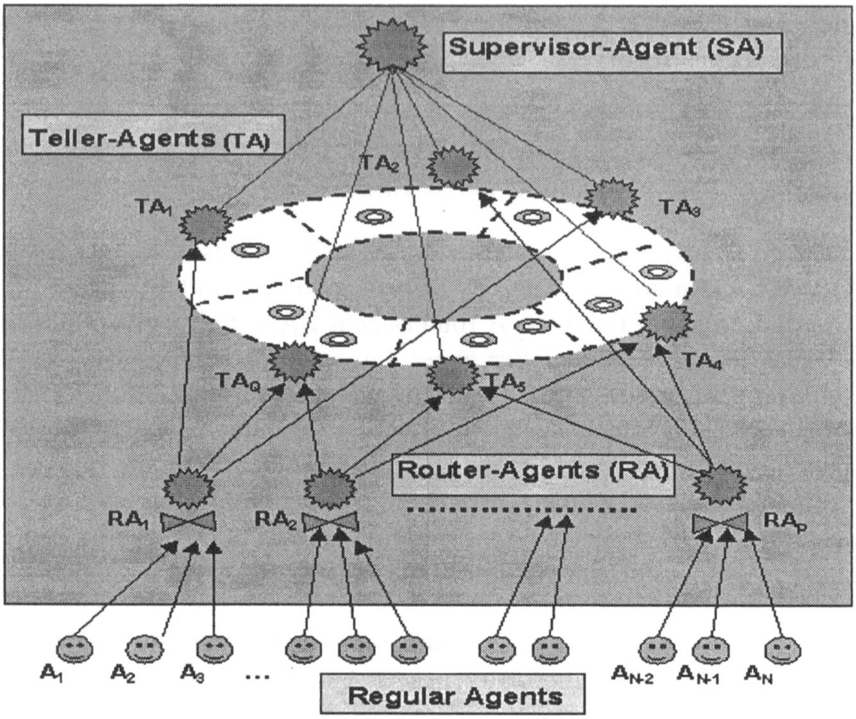

Figure 16. Team Communication Architecture.

7.2.4 Design Agents

Assume that the given IAS is carried out by a supervisor-agent SA, M teller-agents $\{TA_1, TA_2, \ldots, TA_M\}$, and L router-agents $\{ RA_1, RA_2, \ldots, RA_L\}$. Structures and functions of these middle agents can be described as follows.

Supervisor-Agent

Supervisor-agent has the following functions: (i) define the number of channels given to each teller-agent; and (ii) create or destroy teller-agents or router-agents when it is needed. For (i), the supervisor-agent first defines the total number of communication channels Cmax which can be supported by the available resources in the system as the following:

$$C_{\max} = Fc(D_{CPU}, V_M) - A_{\max}$$

Where, Fc is a function, D_{CPU} is the number of threads, and V_M is the volume of memory, given for communication; Amax is the number of channels given to asynchronous mode.

Assume that K_{TAi} is the number of communication channels given to service category Si which is managed by teller-agent TAi, i=1..M. Then, we have:

$$C_{\max} = \sum_{i=1}^{M} K_{TAi}$$

Assume that Wi is the number of requests concerning Si and received by teller-agent TAi. In order to distribute C_{max} channels fairly among M teller-agents, which represent sectors of the rotary dining table, we use the following formula:

$$K_{TAi} = \frac{C_{\max} \times W_i}{\sum_{i=1}^{M} W_i}$$

For (ii), based on the capability of a router-agent, which is the number of requests (CAr) that a router-agent can serve, the supervisor-agent define how many router-agents are needed:

$$L = \frac{N}{CAr}$$

If there are *nR* requests to come then the supervisor-agent creates *nL* new router-agents by the same way:

$$nL = \frac{nR}{CAr}$$

If there are *fR* requests to finish then the supervisor-agent picks:

$$fL = \frac{fR}{CAr}$$

least-loaded router-agents and blocked *(fL-1)* of them so that regular agents can not send their requests to them further. After the requests currently held by them are redirected to the teller-agents, these router-agents become free of requests and therefore can be destroyed without affecting the system.

The control of the number of teller-agents is even simpler as this number is equal to M - the number of service categories in the system. If a new service appears which can not be matched to any of the existing teller-agents then the supervisor-agent creates a new teller-agent for this service. If a teller-agent become idle, i.e. no request, then the supervisor-agent may make it "sleep" instead of destroy it completely as its creation is time consuming.

Router-Agents

The main function of router-agents is to catch requests from regular agents and redirect them to teller-agents based on their service categories or to deliver message to a target receiver agent if such one is specified. As main responsibilities of router-agents are communication-oriented their activities will be described in Section 7.2.5.

Teller-Agent with Rotary-Table

We have M middle teller-agents $\{TA_1, TA_2, \ldots, TA_M\}$, which are responsible for generating and activate communication channels between the agents interested in the same category. The way teller-agents match agent requests is similar to the way a rotary dining table in Chinese restaurants works. Each teller-agent TAj serves a sector Vj of the rotary table which is specialized on a service category Sj. It divides agent requests of the given category into food providers or P-requests and consumers or C-requests. F-requests forms FR(j) queue and simulates the plates in the table. C-requests forms CR(j) queue and simulates the chairs around the table. Thus, in each sector Vj of the rotary table, which is taken care by a teller-agent TAj, we have:

- $PR(j)=\{\{A^1, t_j^1\}, \{A^2, t_j^2\}, \ldots \{A^{u(j)}, t_j^{Uj}\}\}$
- $CR(j)=\{\{A^{1*}, t_j^{1*}\}, \{A^{2*}, t_j^{2*}\}, \ldots \{A^{u(j)*}, t_j^{Uj*}\}\}$

Where, $A^k \in AS=\{A_1, A_2, \ldots, A_E\}$, k=1…Uj or k=1…Uj*; AS is the set of agents who use synchronous mode; t_j^k is the time constraint for the given agent request concerning service Sj, either in providing or demanding. Receiving a request from a router-agent, the teller-agent adds this request to queue PR(j), on the "table," or in queue CR(j), to the "chairs".

Unlike in other communication mechanisms, communication channels in this model are established with consideration of agent time constraints. For a sector Vj with queue FR(j) and CR(j), j=1..M, K_{TAj} channels is given to agents who have shortest life times. The orders of agent requests in queues FR(j) and CR(j) are defined by their priorities which are calculated as the following:

$$PoPR_j^k = G(Ta^k, t_j^k) + Age(Ti) \quad (8)$$

$$PoCR_j^l = H(Ta^l, t_j^l) + Age(Ti) \quad (9)$$

where,

- *G* and *H* are some functions;
- Ta^k or Ta^l are life time of agent A^k or A^k;
- t^k_j and t^l_j are deadlines of agent A^k or A^l interest in Sj;
- *Age(Ti)* is an aging function which increases priority of an agent based on the amount of time the agent is in the system. We use this technique to avoid agent starvation.

Thus, there is no waiting, any new request could be added immediately to the two active queues. Depending on its deadline the communication channel is granted soon or later. In order to match requests in PR(j) with size Uj and requests in CR(j) with size Uj* we use a mechanism which is similar to the hand-shaking process of the two delegations as shown in Figure 17 . The orders of requests are defined using (8) and (9).

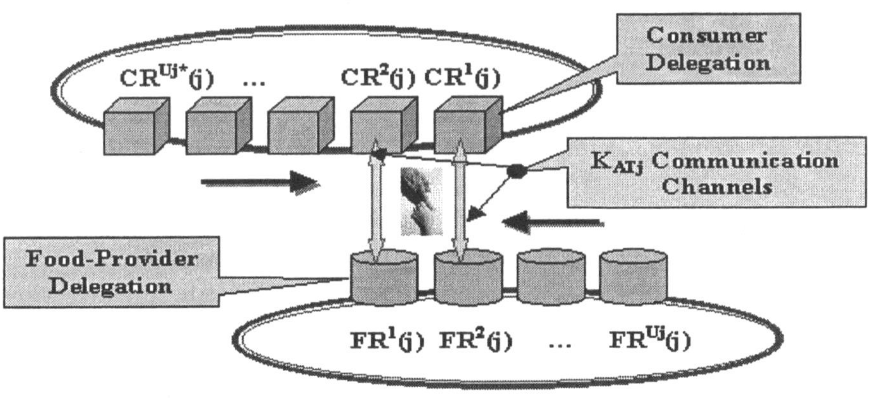

Figure 17. Hand-Shaking Mechanism.

If $K_{TAj} > Max(Uj,Uj^*)$ then it takes:

$$Max\,(Uj\,,Uj^*) + Min\,(Uj\,,Uj^*) - 1$$

cycles for all requests in this category to "talk" with all other possible partners. If $K_{TAj} < Max(Uj,Uj^*)$ then it takes:

$$\frac{(Max\ (Uj\,,Uj\,^*) + K_{TAj} - 1) \times Min\ (Uj\,,Uj\,^*)}{K_{TAj}}$$

cycles for all requests in this category to "check" with all other possible partners, compared with *n-1* cycles using Yellow-page or Contract-Net methods. Recall that n >> Uj and Uj*.

7.2.5 Build Communication System

Communication protocols managed by Teller-Agents and Router-Agents can be illustrated as in Figure 18. In order to register, an agent Ai sends a message with command "O1:Start" in the COM. A router-agent catches it and creates a message box and sets the pointer to the given agent. Since then, any message addressed the given agent will be automatically forwarded to this personal box which is deleted only when the given agent send a message with "O8:Finish" in COM.

In asynchronous mode, an agent A_i uses COM=Send_A(A_y) to send a message directly to a target agent A_y at any time when it needs. If the router-agent receives such a command in the message, it delivers this message into the receiver A_y's message box. Agent A_y can reply by the same way with COM=Send_A(A_i).

In synchronous mode, an agent can use services of the teller-agents to create a communication channel to agents who would meet his interests. In this case the given agent should use COM=Open and put the list of interests in the BODY; If the router-agent receives such a command in the message, it checks the BODY, extracts it into requests: one service category, exp. Sk per request, and then redirects messages to teller-agents, exp TA_k which is responsible for the given service category Sk.

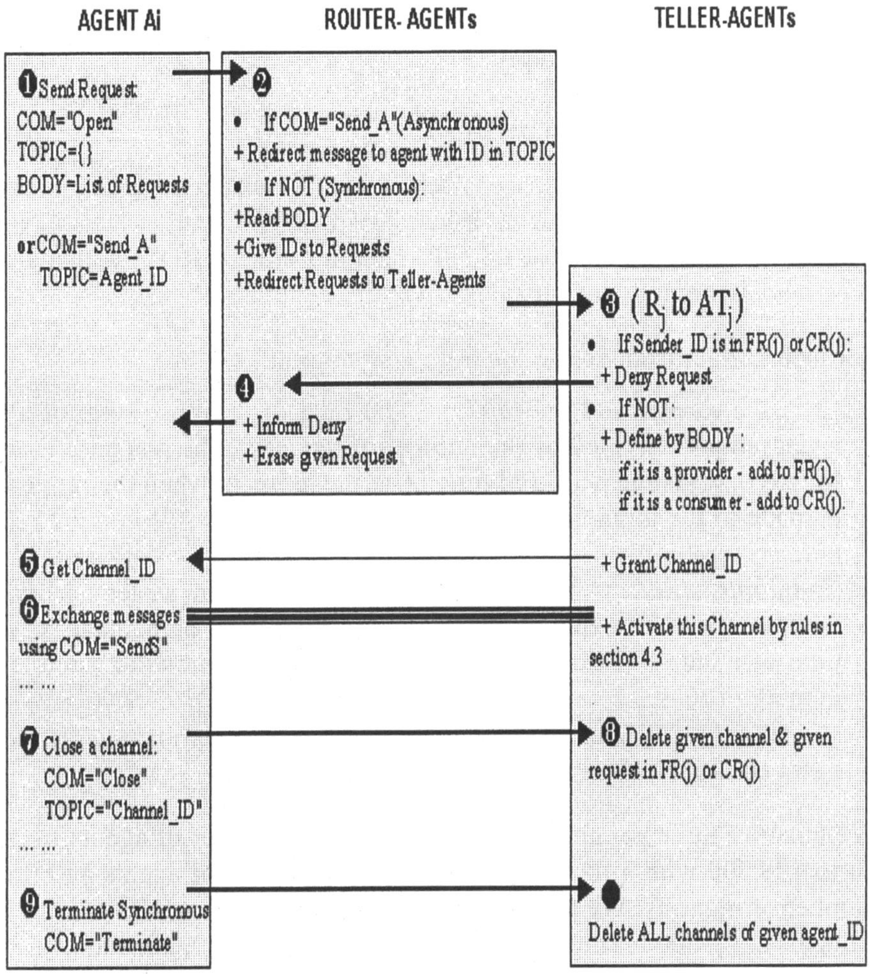

Figure 18. Communication Protocol.

7.2.6 Coordinate Components

In this IAS the trading processes are divided into transactions each of which has a buyer, a seller and is for one kind of good or service. Thus, an agent may be involved in a number of transactions. In coordinating these agents we need to use the strict order method with a global timing system to guarantee that agents who make bids first will have their requests to be proceeded first. Since data in this IAS will be shared by many components and agents in a large-scale fashion, in order to avoid the bottleneck problem we need to replicate data and use the Strict Consistency model (Gali 2000).

7.2.7 Analyze and Evaluate IAS

For the same e-marketplace scenario we use five different communication methods: Yellow-pages or Black-board (YP), Contract-Net (CN), Pattern-based (PB), Peer-to-Peer (PP), and our model (RDT) which are described in previous sections.

For their comparison we use the following criteria:

- *Average Response Time (aRT):* is the total of response time of all agent requests divided by the number of requests; The response time for a request is the period of time from the moment when the request is made to the moment when it is known if it can result in a deal or not.
- *Reliability (Re):* is the probability that the given system works without crash, in experiments we can define it in percentage as a ratio of number of times the system work without crash to the total number of runs;
- *Communication Degree (cD):* number of channels created in the agent communication system at a time.

We measure these parameters while varying the number of agents from 10 to 100,000 with a fixed ratio between the number of agents and the number of requests. We assume that the system capacity is fixed with a value of C_{max}=5000. The results are shown in Figures 19, 20, and 21.

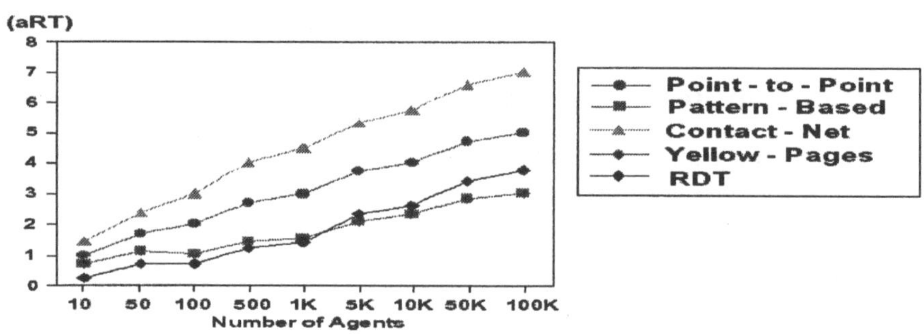

Figure 19. Response Time.

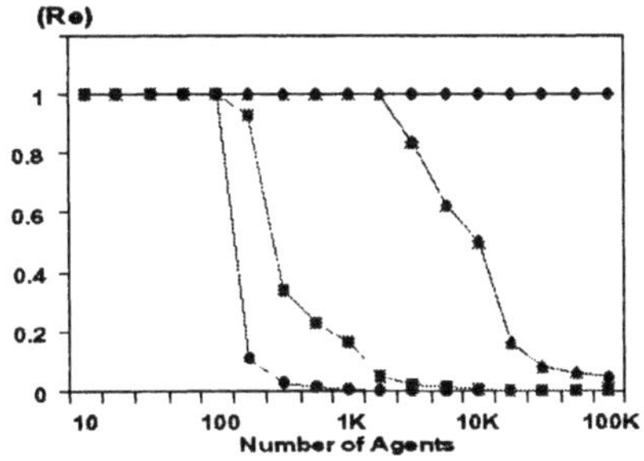

Figure 20. Reliability.

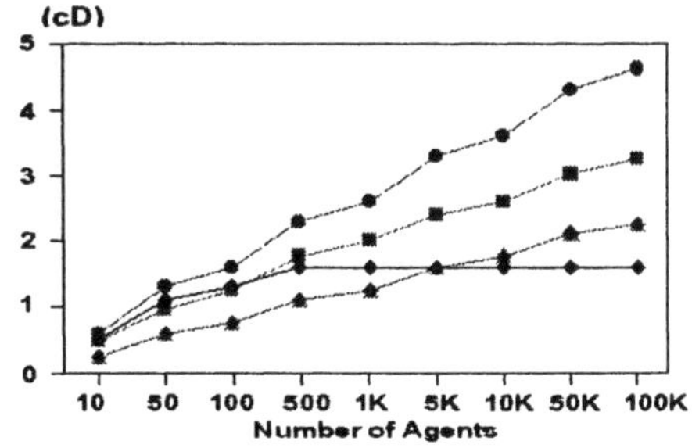

Figure 21. Communication Degree.

The graphics show that RDT, PB, and YP produce good response times and PB and YP are better than RDT. However, by reliability PB is unacceptable. For YP, it is doing well until the number of agents is less than 3,000. However, when starts from 5,000 the reliability dramatically drops, when the number of agents reaches 10,000 it remains only about 0.5, i.e. 50% of runs are failed. After that it comes down to about 0.1 or 90% failure when the number of agents is 20,000. Since the acceptable reliability of a system need to be at least not less than 0.9, up to 100 agents any method would be reliable, PB can be used for systems with up to 300 agents, YP and CN can be used for systems up to 4000, and only RDT can be used for any size, in term of reliability. Thus, RDT can support massive multi-agent systems of large number of agents with good response times and an absolute reliability. The stability of RDT can be explained by graphics in Figure 21 which shows that while the numbers of communication channels created in other methods increase as the number of agents grows, in RDT it remains the same after a point. That is because the number of communication channels in RDT is kept to be not more than C_{max}, thanks to the control done by the team of middle agents, particularly the Supervisor-agent in cooperation with Teller-agents.

Besides, note that in our architecture an agent can be in both synchronous and asynchronous communication modes simultaneously. It gives agents more freedom and flexibility. The asynchronous mode is good for one-to-one negotiation, while synchronous mode would be suitable for surveys. First, agents can use synchronous communication for a brief look at the e-market, having short conversations with the potential agent-partners arranged by teller-agents with the hand-shaking mechanism. After selecting the right partners, the agent can exploit asynchronous mode as they already know the target partners.

8 Conclusions

In this chapter we have described the principles of Internet agent-based systems which is an integration of distributed computing and artificial intelligence. We have introduced and classified Internet agent-based systems, their architectures, design procedure, and their applications. We have shown how software agents can be used to build autonomous and collaborative Internet-based systems and how agents themselves participate in developing and designing these systems. As nowadays many business and engineering systems are large-scale and distributed Internet agent-based systems are perfectly suitable solutions to conduct these processes in an automatic, collaborative, and efficient way.

Acknowledgments

This work is supported in part by the SUNY New Palts Research Creative Award 00-01 and 01-02. Many thanks are due to the editors for their very helpful comments which improve both the presentation and the content of this chapter.

References

Arisha, K., Eiter, T., Kraus, S., Ozcan, F., Ross, R., and Subrahmanian, V. (1999), "Impact: a platform for collaborating agents," *IEEE Intelligent Systems*, vol. 14, no. 2, March/April, pp. 64-72.

Albert, B., Parunak, H., and Erol, K. (1999), "Agents and the Internet: infrastructure for mass customization," *IEEE Internet Computing*, vol. 3, no. 5, September/October, pp. 62-69.

Bakos, Y. (1998), "The emerging role of electronic marketplaces on the Internet," *ACM Communications*, vol. 41, no. 9, September, pp. 35-42.

Beck, M. (1999), "Active and real-time functionality for electronic brokerage design," *1st WECWIS*, pp. 12-22.

Ben-Daya, M. and Rahim, A. (1999), "Multi-stage lot sizing models with imperfect processes and inspection errors," *Production Planning and Control*, vol.10, no. 2, pp. 118-126.

Bradshaw, J., Greaves, M., Holmback, H., Karygiannis, T., Silverman, B., Suri, N., and Wong, A. (1999), "Agents for the masses," *IEEE Intelligent Systems*, vol. 14, no. 2, March/April, pp. 53-63.

Cassandra, A., Chandrasekara, D., and Nodine, M. (2000), "Capability-based agent matchmaking," *Autonomous Agents 2000*, pp. 201-202.

Decker, K., Sycara, K., and Williamson, M. (1997), "Middle Agents for the Internet," Tech. Report at Cora Research Database, pp. 1-6.

Elofson, G. and Robinson, W. (1998), "Creating a custom mass-production channel on the Internet," *ACM Communications*, vol. 41, no. 3, Mar., pp. 56-62.

Evans, E. and Rogers, D. (1997), "Using Java applets and Corba for multi-user distributed applications," *IEEE Internet Computing*, vol. 1, no. 4: July-August, pp. 43-48.

Ferber, J. (1999), *Multi-agent Systems*, Addison-Wesley.

Froehlich, G. (1999), "Application framework issues when evolving business application for e-commerce," *Proceedings of 32nd Hawai Int. Conference on Computer Science*, pp. 1-10.

Gali, D. (2000), *Distributed Operating Systems*, Prentice Hall.

Grecu, D. and Brown, D. (1996), "Learning to design together," *Proceeding of the AAAI-96 Spring Symposium on Adaptation, Coevolution, and Learning in Multi-Agent Systems*, pp. 1-6.

Huhns, M. and Malhotra, A. (1999), "Negotiating for goods and services," *IEEE Internet Computing*, vol. 3, no. 4, July/August, pp. 97-99.

Ignacio, S., Mercedes, G., Carlos, I.A., and Ramos, M. (2000), "An agent architecture to fulfill real-time requirements," *Autonomous Agents 2000*, pp. 475-482.

Jamali, J., Thati, P., and Agha, G. (1999), "An actor-based architecture for customizing and controlling agent ensembles," *IEEE Intelligent Systems*, vol. 14, no. 2, March/April, pp. 38-44.

Janin, K. and Singh, M. (1999), "Agents for process coherence in virtual enterprises," *ACM Communication*, vol. 42, no. 3, pp. 62-80.

Khare, R. and Rifkin, A. (1997), "XML: a door to automated web applications," *IEEE Internet Computing*, vol. 1, no. 4, July-August, pp. 78-87.

Kiniry, J. and Zimmerman, D. (1997), "A hands-on look at Java mobile agents," *IEEE Internet Computing*, vol. 1, no. 4, July-August, pp. 21-30.

Knapik, M. and Johnson, J. (1998), *Developing Intelligent Agents for Distributed Systems*, McGraw-Hill.

Lange, D. and Oshima, M. (1999), "Seven good reasons for mobile agents," *ACM Communication*, vol. 42, issue 3, pp. 88-89.

Mammen, D. and Lesser, V. (1997), "A testbed for the evaluation of multi-agent communication and problem-solving strategies," *Cora Research Database*, pp. 1-6.

McFall, C. (1998), "An object infrastructure for Internet middleware IBM on component broker," *IEEE Internet Computing*, March-April, pp. 46-51.

Ma, M. (1999), "Agents in e-commerce," *ACM Communications*, vol. 42, no. 3, pp. 79-80.

Nardi B., Miller J., and Wright D. (1998), "Collaborative Programmable Intelligent Agents," *ACM Communications*, Vol.41, N3, pp. 99-104.

Nissen M and Mehra A. (1999), "Some Intelligent Supply Chain Agents," *Proceeding of Autonomous Agents'99*, 374-375.

Nwana H., Ndumu D., Lee L. (1998), "ZEUS: An Advanced Tool-Kit for Engineering Distributed Multi-Agent Systems," *BT Technical Report*, pp. 1-15.

Dabka P. (1999), "Enterprise Integration via COBRA-based Information Agents," *IEEE Internet Computing*, Vol. 3, No. 5, September/October, pp. 49-57.

Parunak H, Baker A., Clark S. (1998), "The AARIA Agent Architecture: From Manufacturing Requirements to Agent-Based System Design," Proceeding of the *Workshop on Agent-based Manufacturing* '98, pp. 1-17.

Pfeiffer, T. (1999), "Transfer Pricing and Decentralized Dynamic Lot-sizing in Multistage, Multiproduct Production Processes". *European Journal of Operational Research*, N.116, pp. 319-330.

Pham H. and Nguyen H. (2001), "Agent Genders for Inventory Management," in *Lecture Notes in Artificial Intelligence* (LNAI), Springer-Verlag Pub., (in Press).

Pham H., Nguyen H., Nguyen V. (2000), "Environment and Means for Cooperation and Interaction in E-commerce Agent-based Systems," proceeding of the *International Conference on Internet Computing* (IC'2000), pp. 253-259.

Pham H., Simonenko V. (1997), "Objective-Oriented Algorithm for Job Scheduling in Parallel Heterogeneous Systems," in *Lecture Notes in Computer Science* (LNCS), Springer-Verlag Pub., Vol. 1291, pp. 193-214.

Pham H., Tran S. (2000), "Round-Table Architecture for Communication in Multi-Agent Softbot Systems," in *Lecture*

Notes in Computer Science, Springer-Verlag Pub. Vol. 1983, pp. 338-344.

Pham H., Worasing R. , Thanadee U. (1999), "Distributed Multi-level Adaptation for Dynamic Multi-Agent Systems," in the Proceeding of *22nd IEEE Conference on Systems, Man, and Cybernetics*, pp. 1-6.

Rana Omer. (2000), "Performance management of mobile agent system," *Autonomous Agents'2000*, pp. 148 - 155.

Rana Omer., Stout K. (2000), "What is Scalability in Multi-Agent Systems," *Autonomous Agents'2000*, pp. 56-63.

Reeves, D., Grosof, B., Wellman, M., and Chan, H. (1999), "Toward a Declarative languages for Negotiating Executable Contracts," *AAAI Workshop on AI for E-commerce*, pp. 39-45.

Schmidt, D. (1998), "Evaluating architectures for multithreaded object request brokers," *ACM Communications*, vol. 41, no. 10, pp. 54-60.

Senn, J. (1999), "The Evolution of B2B Commerce Models," *1st WECWIS*, pp. 1-10.

Shehory, O. and Sycara, K. (2000), "The RETSINA communicator," *Autonomous Agents 2000*, pp. 199-200.

Shen, W. and Norrie, D. (1999), "Developing Intelligent Manufacturing Systems using Collaborative Agents," Proceeding of the *Workshop on Intelligent Manufacturing Systems*, pp. 1-8.

Shu, S. and Norrie, D. (1999), "Patterns for Adaptive Multi-agent Systems in Intelligent Manufacturing," Technical Report of Univ. of Calgary, pp. 1-8.

Smith, R.G. (1980), "The contract Net protocol: High-level Interaction and Control in a Distributed problem Solver," *IEEE Tran. On Computers*, V29-N12, pp. 1104-1113.

Suri, N. and Bradshaw, J. (2000), "NOMADS: toward a strong and safe mobile agent system," *Autonomous Agents'2000*, pp. 163-164.

Wong, D., Paciorek, N., and Moore, D. (1999), "Java-based mobile agents," *ACM Communications*, vol. 42, No. 3, pp. 92-105.

Yamamoto, G. and Nakamura, Y. (1999), "Architecture and performance evaluation of a massive multi-agent system," *Autonomous Agents '99*, pp. 319-325.

Yamamoto G. and Tai H. (2000), "Architecture of an agent server capable of hosting tens of thousands of agents," *Autonomous Agents'2000*, pp. 70 - 71.

Labrou Y., Finin T., and Peng Y.(1999), "Agent Interaction Languages: The Current Landscape," *IEEE Intelligent Systems*, Vol. 14, No. 2, March/April, pp. 45-52.

Chapter 6

Compositional Design and Maintenance of Broker Agents

C.M. Jonker and J. Treur

In this chapter a generic broker agent architecture is introduced and designed in a principled manner using the compositional development method for multi-agent systems DESIRE. A flexible and easily adaptable agent architecture is implemented including automated support of the agents own maintenance. Therefore, the agent is not only easily adaptable, but it shows its adaptive behaviour to meet new requirements, supported by communication with a maintenance agent.

1 Introduction

To support users on the World Wide Web, various types of agents can be, and actually have been, developed. For example, to support brokering processes in electronic commerce, personal assistant agents can be developed that support a user offering products (or services) at the Web, or agents that support search for information on products within a user's scope of interest, or agents that combine both functionalities. Moreover, mediating agents can be developed that communicate both with personal assistant agents that provide information on products and with personal assistants that ask for information on products. Recently a few applications of broker agents have been addressed for this area [for example, see (Chavez *et al.* 1996, Chavez *et al.* 1997, Kuokka and Harada 1995, Martin *et al.* 1997, Sandholm and Lesser 1995, and Tsvetovatyy and Gini 1996)]. In general, applications like these are implemented in an ad hoc fashion without an explicit design at a conceptual level, and without taking into account the dynamic requirements imposed by the domain of application and the maintenance problem (including desired extension or modification of functionality) implied by this dynamic character.

In this chapter a generic agent architecture for agents acting in brokering processes is introduced which has been designed in a principled manner, using the compositional development method for multi-agent systems DESIRE. The agent architecture can be instantiated by adding specific types of knowledge to support functionalities and behaviour required. Depending on the choice of these requirements, an agent is created for a specific context by including the appropriate types of knowledge. For example, a search agent with functionality restricted to (incidental) search for information upon a user's request can be built by adding only knowledge needed for this task. Such an agent, for example, is not able to store and maintain the user's query or information that has been found, nor is it able to provide information to other agents. If these functionalities are required as well, the necessary types of knowledge have to be added.

Requirements imposed at the time of creation determine the functionalities and behaviour of the agent. If later on additional or changed requirements come up, the agent has to be modified or redesigned to meet the new requirements. The agent architecture introduced here supports its own modification due to the fact that basic functionalities are specified in an explicit declarative manner, in the form of knowledge. It is possible to dynamically modify the agent by adding or deleting some of its knowledge. Since this declaratively represented knowledge can be the subject of communication between agents, modification can be performed at a distance: another agent (e.g., a dedicated maintenance agent) communicates the knowledge needed for the modification to the agent that is to be modified. Thus a flexible agent implementation is obtained that can be maintained and evolve over time on the basis of communication with other agents only.

In Section 2 an example problem domain for brokering processes is sketched. Section 3 introduces the design of the generic agent architecture. The different types of knowledge are presented in Section 4. In Section 5 the behaviour of the system is analysed by giving an overview of which types of knowledge are needed for which types of basic functionalities. Finally, in Section 6 it is shown how the agent can communicate that a specific functionality is required, and it is explained how, after appropriate maintenance knowledge has been communicated to the agent, the agent performs its own dynamic modification in order to obtain the required functionalities.

2 Electronic Commerce and Brokering

The process of brokering as often occurs in electronic commerce involves a number of agents. A user offering products may be supported by a provider agent that provides information about the products to other (human or computer) agents. A user looking for products may be supported by a personal assistant agent that takes its user's queries and contacts other agents or looks at the Web directly to find information on products within the user's scope of interest. Such a personal assistant agent may contact either provider agents immediately, or mediating agents, which in turn have contact with provider agents, or other mediating agents. Depending on the application, the chain of agents involved may include zero or more mediating agents.

The domain analysed for the agent architecture presented here is the domain of brokering (scientific) papers. This domain has a number of aspects in common with other domains within the area of electronic commerce. The task of a provider agent is to inform other researchers on papers available on Internet (a marketing aspect); for example an agent related to a Web site of a research group, announces new papers included in their Web site. If a researcher is looking for a paper with certain characteristics (scope), a personal assistant agent can ask other agents for information on papers with these characteristics. To be able to tune the information provided to users, a number of scopes of interest can be maintained for each of the users. For example, one of the users may be interested in papers on certain topics, such as work flow management systems, but also in papers on agents and the World Wide Web.

Topics can be basic (e.g., 'work flow management systems', or 'agents', or 'World Wide Web'), or a combination of a number of topics (e.g., 'agents and World Wide Web'). In the latter case the user interest is limited to papers which address both topics. Moreover, if it is added that the user is only interested in papers from the years 1995 to 1997, then either year in the range 1995-1997 is meant. Topics can be matched with, for instance, the set of keywords of a paper, or with the abstract, or the paper as a whole. In some disciplines, such as Medicine, an ontology of topics has been developed that serves more or less as a standard. Besides topics also other attributes of papers can be used to define a scope of interest, for example an author, a year, a research group, et cetera. These attributes can also be used in combination with each other. For this example, a shared ontology of topics is assumed. All agents in the brokering process express their information and interests using this shared ontology. It is assumed that the following attributes of a paper are available and can be used: title, authors, affiliation(s) of the authors, location on the World Wide Web where it can be found, topics covered by the paper, abstract, year, and reference. This information can be used to identify papers that are of interest to a user, but also forms the source for the information that can be provided to a user when a paper is proposed to him or her.

3 Compositional Design of the Generic Broker Agent

3.1 Compositional Design of Multi-Agent Systems

The emphasis in the compositional design method for multi-agent systems DESIRE (cf., [1]) is on the conceptual and detailed design. The design of a multi-agent system in DESIRE is supported by graphical design tools within the DESIRE software environment. The software environment includes implementation generators with which (formal) design specifications can be translated into executable code of a prototype system. In DESIRE, a design consists of knowledge of the following three types: *process composition*, *knowledge composition*, and the *relation* between process composition and knowledge composition. These three types of knowledge are discussed in more detail below.

3.1.1 Process Composition

Process composition identifies the relevant processes at different levels of (process) abstraction, and describes how a process can be defined in terms of (is composed of) lower level processes.

Identification of Processes at Different Levels of Abstraction

Processes can be described at different levels of abstraction; for example, the process of the multi-agent system as a whole, processes defined by individual agents and the external world, and processes defined by task-related components of individual agents. The identified processes are modelled as *components*. For each process the *input and output information types* are modelled. The identified levels of process abstraction are modelled as *abstraction/specialisation relations* between components: components may be *composed* of other components or they may be *primitive*. Primitive components may be either reasoning components (i.e., based on a knowledge base), or, components capable of performing tasks such as calculation, information retrieval, optimisation. These levels of process abstraction provide process hiding at each level.

Composition of Processes

The way in which processes at one level of abstraction are composed of processes at the adjacent lower abstraction level is called *composition*. This composition of processes is described by a specification of the possibilities for *information exchange* between processes (*static view* on the composition), and a specification of *task control knowledge* used to control processes and information exchange (*dynamic view* on the composition).

3.1.2 Knowledge Composition

Knowledge composition identifies the knowledge structures at different levels of (knowledge) abstraction, and describes how a knowledge structure can be defined in terms of lower level knowledge structures. The knowledge abstraction levels may correspond to the process abstraction levels, but this is often not the case.

Identification of knowledge structures at different abstraction levels

The two main structures used as building blocks to model knowledge are: *information types* and *knowledge bases*. Knowledge structures can be identified and described at different levels of abstraction. At higher levels details can be hidden. An *information type* defines an ontology (lexicon, vocabulary) to describe objects or terms, their sorts, and the relations or functions that can be defined on these objects. Information types can logically be represented in order-sorted predicate logic. A *knowledge base* defines a part of the knowledge that is used in one or more of the processes. Knowledge is represented by formulae in order-sorted predicate logic, which can be normalised by a standard transformation into rules.

Composition of Knowledge Structures

Information types can be composed of more specific information types, following the principle of compositionality discussed above. Similarly, knowledge bases can be composed of more specific knowledge bases. The compositional structure is based on the different levels of knowledge abstraction distinguished, and results in information and knowledge hiding.

3.1.3 Relation between Process Composition and Knowledge Composition

Each process in a process composition uses knowledge structures. Which knowledge structures are used for which processes is defined by the relation between process composition and knowledge composition.

3.2 Design of the Generic Broker Agent

For the design of the generic broker agent the main aspects introduced above are considered: process composition, knowledge composition, and relations between knowledge and process composition. A compositional generic agent model GAM (introduced in [3]), supporting the weak agency notion (cf. [13]) is used. At the highest abstraction level within an agent, a number of processes can be distinguished that support interaction with the other agents. First, a process that manages communication with other agents, modelled by the component agent interaction management in Figure 1. This component analyses incoming information and determines which other processes within the agent need the communicated information. Moreover, outgoing communication is prepared. Next, the agent needs to maintain information on the other agents with which it co-operates: maintenance of agent information. The component maintenance of world information is included to store the information on world information (e.g., information on attributes of products). The process own process control defines different characteristics of the agent and determines foci for behaviour. The component world interaction management is included to model interaction with the world (with the World Wide Web world, in the example application domain): initiating observations and receiving observation results.

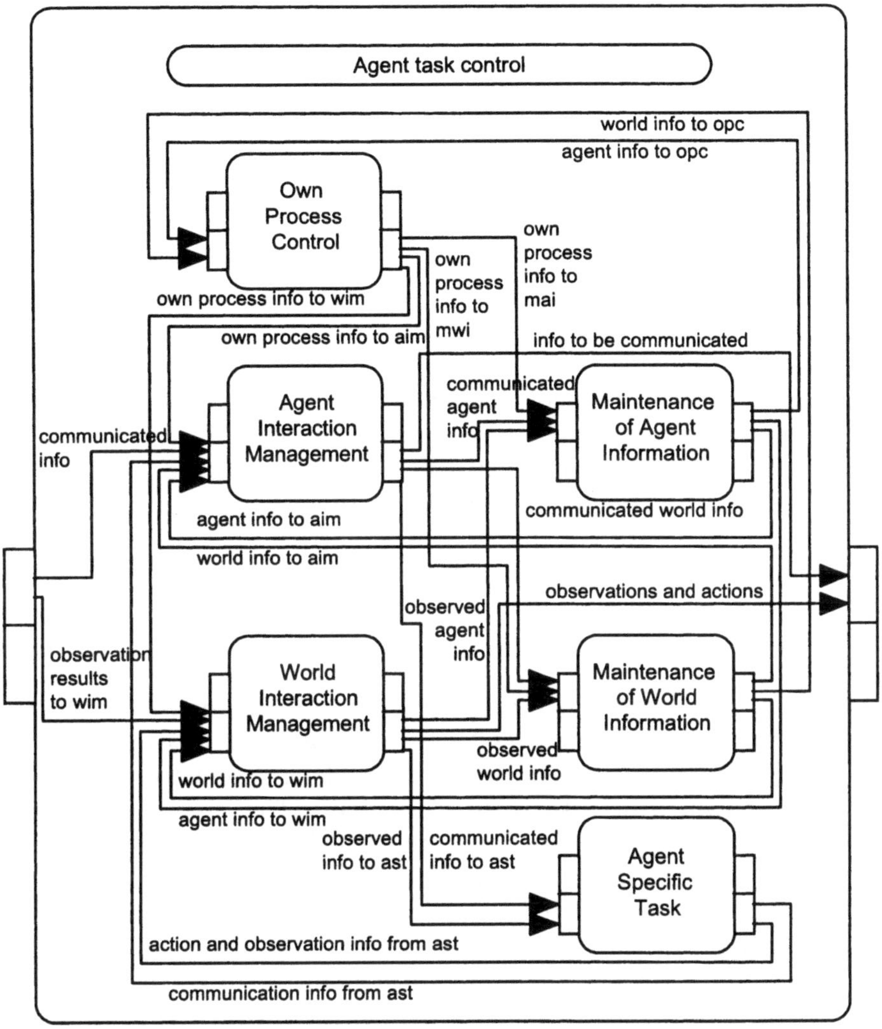

Figure 1. Composition at the highest level within the generic agent model GAM.

The agent processes discussed above are generic agent processes. Many agents perform these processes. In addition, often agent-specific processes are needed: to perform tasks specific to one agent, for example directly related to a specific domain of application. In the current example the agent has to determine proposals for other agents. In this process information on available products (communicated by information providing agents and kept in the component maintenance of world information), and about the scopes of interests of agents (kept in the component maintenance of agent information), is combined to determine which agents might be interested in which products. For the broker agent this agent-specific task is called determine proposals. Figure 1 depicts how the broker agent is composed of its components (process composition); this composition is simply reused from the generic agent model GAM.

Part of the exchange of information within the generic agent model GAM can be described as follows. The broker agent needs input about scopes of interests put forward by agents and information about attributes of available products that are communicated by information providing agents. It produces output for other agents about proposed products and the attributes of these products. Moreover, it produces output for information providers about interests. In the information types that express communication information, the subject information of the communication and the agent from or to whom the communication is directed are expressed. This means that communication information consists of statements about the subject statements that are communicated.

Within the broker agent, the component own process control uses belief input information and generates focus information: to focus on a scope of interest to be given a preferential treatment, i.e., pro-active behaviour will be shown with respect to this focus. The component agent interaction management has the same input information as the agent (incoming communication), extended with belief info and focus info. The output generated includes part of the output for the agent as a whole (outgoing communication), extended with maintenance info (information on the world and other agents that is to be stored within the agent), which is used to prepare the storage of communicated world and agent information.

Information on attributes of products is stored in the component maintenance of world information. In the same manner, the beliefs of the agent with respect to other agents' profiles (provider attribute info and interests) are stored in maintenance of agent information. The agent specific task determine proposal uses information on product attributes and agent interests as input to generate proposals as output.

For reasons of space limitation the generic and domain-specific information types within the agent model are not presented.

4 Generic and Domain Specific Knowledge

The different knowledge abstraction levels introduced for information types can also be exploited to structure the knowledge. Parts of the knowledge can be formulated in terms of scopes, abstracting from attributes and values. Other knowledge is used to perform the abstraction step: it can be used to derive conclusions in terms of scopes from input in terms of attributes and values. The knowledge bases are discussed below in the context of the component in which they are used.

4.1 Agent Specific Task: Determine Proposals

To determine proposals fitting a given scope of interest, information on products has to be compared to this scope of interest. To this end, the information on products, expressed in terms of their attributes has to be aggregated to information in terms of scopes. This can be derived using two knowledge bases, attribute and scope kb, which defines the relations between attributes and scopes in general, and product scope abstraction kb, which identifies for which scope(s) a product is relevant.

The composition of the knowledge in these two knowledge bases supports reuse. For example, if in one of the two knowledge bases, modifications are made, the other knowledge base still can be used. Moreover, the first knowledge base is specified independent of knowledge about products. It can be (re)used within the component maintenance of agent information as well, as will be shown below.

Given information on the scopes of products, by the knowledge base strict match kb it is defined how proposals to agents can be generated by matching the scopes of products and the scopes in which an agent is interested. For strict matching it consists of only the following knowledge.

if A is interested in scope S
and product P is in scope S
then product P is interesting for A because of scope S

In formal specification form this is expressed as:

Knowledge base strict match kb
if interested_in(a:agent, s:scope)
and in_scope(p:product, s:scope)
then is_interesting_for(p:product, a:agent, s:scope)

This knowledge simply states that if a product is in a scope an agent is interested in, then this product is interesting for this agent. Alternative knowledge bases can be used for non-strict matching.

4.2 Agent Interaction Management

The component agent interaction management makes use of four knowledge bases: (1) for incoming communication from agents asking for information on products, (2) incoming communication from agents providing information, (3) outgoing communication to agents interested in information on products, and (4) outgoing communication to agents providing information.

4.2.1 Incoming Communication

If an agent communicates her or his interests, then this information is identified as new agent interest information that is currently believed (which can be forgotten after the agent has reacted on it: knowledge base agent interest identification kb) or that has to be stored (in which case it can be remembered later: knowledge base agent interest maintenance identification kb). A condition for storage of interests information is that the type of contract is persistent. For agents with a weaker type of contract no requests are stored. The knowledge used in structured natural language is:

if communicated by agent A: interest in scope S
and belief: A has a contract of type persistent
then new agent info: A is interested in scope S

In concise formal form this is specified as:

Knowledge base agent interest maintenance identification kb

if communicated_by(interest(s:scope), v:sign, a:agent)
and belief(has_contract(a:agent, contract_type(persistent,y)), pos)
then new_agent_info(interested_in(a:agent, s:scope), v:sign)

If an agent communicates that he or she wants to subscribe for a contract of a certain type, then this information is identified as new contract information that has to be stored. This identification makes use of the following knowledge.

if communicated by agent A: subscription for contract type C
then new agent info: A has contract of type C

In formal form:

Knowledge base subscription identification kb

if communicated_by(subscription_for(c:contract_type), v:sign, a:agent)
then new_agent_info(has_contract(a:agent, c:contract_type), v:sign)

If an agent communicates information about products it provides, this incoming information is analysed, new world information is identified as new information that can be used immediately and forgotten afterwards (knowledge base provider info identification kb), or has to be stored (knowledge base provider info maintenance identification kb). If an agent communicates information about products it provides, this incoming information can also be analysed, in order to obtain new agent information on the scopes of the information the agent (apparantly) can provide. This is expressed by:

if communicated by agent A:product P's attribute A has value V
then new agent info:agent A can provide attribute A with value V

In concise formal form:

Knowledge base provider scope maintenance identification kb:

if communicated_by(attribute_has_value(p:product, a:attribute, v:value), pos, a:agent)
then new_agent_info(can_provide(a:info_provider, a:attribute, v:value), pos)

4.2.2 Outgoing Communication

New information (product identification, scope, or attribute information) on a product that may be interesting for an agent is communicated to this agent. This is expressed in the following knowledge.

if belief: product P is interesting for agent A because of scope S
and belief: I is info on product P (of the form P's attribute A has value V)
then is to be communicated to A: P is interesting because of scope S
and is to be communicated to A: info I

In concise formal form:

Knowledge base proposal communication kb:
if belief(is_interesting_for(p:product, a:agent, s:scope), pos)
and belief(attribute_has_value(p:product, a:attribute, v:value), pos)
then to_be_communicated _to(is_interesting(p:product, s:scope), pos, a:agent)
and to_be_communicated _to(attribute_has_value(p:product, a:attribute, v:value), pos, a:agent)

The agent only communicates to an information provider if a scope has been taken as a focus, and if the information provider can provide products on this scope. This is expressed by:

if in search focus: scope S
and belief: agent A can provide scope S
then is to be communicated to agent A: interest in scope S

In concise formal form:

Knowledge base info provider request kb
if in_search_focus(s:scope)
and belief(can_provide_scope(a:agent, s:scope), pos)
then to_be_communicated _to(interest(s:scope), pos, a:agent)

4.3 Own Process Control

By means of the knowledge base focus kb, used within own process control, the types of proposals to be determined and the scopes on which to focus search are determined, as indicated by the following knowledge base. For example, in the first knowledge element it is expressed that for an agent with a contract of any type, proposals will be determined that fit the agent's interests. This is in contrast with, for example, the second knowledge element which expresses that only for agents with a persistent contract type, their scopes of interests will be chosen as persisting search foci (otherwise these scopes of interest will be forgotten after being handled).

Knowledge base focus kb

if belief(has_contract(a:agent, c:contract:type))
and belief(interested_in (a:agent, s:scope), pos)
then product_to_be determined(in_scope (p:product, s:scope))
and proposal_to_be determined(is_interesting_for(p:product, a:agent, s:scope))

if belief(has_contract(a:agent, contract_type(persisting, search_for_info)))
and belief(interested_in (a:agent, s:scope), pos)
then in_persisting_search_focus(s:scope)

if belief(has_contract(a:agent, contract_type(incidental, search_for_info)))
and belief(interested_in (a:agent, s:scope), pos)
and **not** search_focus_processed_for(s:scope, a:agent)
then in_incidental_search_focus(s:scope)
and search_focus_chosen_for(s:scope, a:agent)

if in_persistent_search_focus(s:scope)
then in_search_focus(s:scope)

if in_incidental_search_focus(s:scope)
then in_search_focus(s:scope)

if in_search_focus(s:scope)
then provider_to_be_determined_for(s:scope)

4.4 World Interaction Management

The component world interaction management alllows the agent to look for information by observation. This entails generation of observations to be performed, and obtaining the observation results. The obtained observation results can be used incidentally after which the information is forgotten (using knowledge base observation info identification kb) or maintained to be used later as well (using knowledge base observation maintenance identification kb, similar to agent interaction management. The agent only observes if a scope has been taken as a focus. This is expressed using knowledge base observation initiative kb.

Knowledge base observation initiative kb
if in_search_focus(s:scope)
then to_be_observed(s:scope)

The actual execution of the observation does not take place within the agent, but in the external world. As part of the external world an engine can be used to search for products matching the pattern defined by the specified scope. The result of such an observation will be all information of any product that matches the scope. The knowledge base specified above is kept rather simple. To avoid too frequent repetition of observation, more sophisticated knowledge can be specified.

4.5 Maintenance of World and Maintenance of Agent Information

In principle, the components maintenance of world information and maintenance of agent information store information. The knowledge base attribute and scope kb defined above is also used in the component maintenance of agent information . In addition, the knowledge base provider scope abstraction kb is defined; it is similar to the knowledge base product scope abstraction kb mentioned above.

5 The Behavior

The behavior of the broker agent can be analysed in different ways. One way is to consider its basic functionalities with respect to its brokering task, and use these as building blocks to obtain behaviour. For example, its behaviour in terms of the weak notions of agency (autonomy, social ability, reactivity, and pro-activity) can be determined in terms of basic functionalities. Moreover, basic functionalities can be related to knowledge bases that are available within the agent. Using these two relationships, a relation can be identified between behaviour and available knowledge within the agent.

5.1 Basic Functionalities Depending on the Agent's Knowledge

The broker agent shows behaviour depending on certain basic functionalities. For the agent model presented, these basic functionalities have been specified in a declarative manner by the agent's knowledge. For each of the basic properties of the agent it has been established which knowledge bases are required. By varying the choice of knowledge for the agent, different types of agents can be designed.

1. *Observation of information available within a certain part of the world*

Observation requires the ability to iniate observations, specified in the knowledge base observation initiative kb, and the ability to identify the information resulting from an observation, specified in the knowledge base observation info identification kb. Both knowledge bases can be used within the component world interaction management.

2. *Communication with agents asking for information on products*

The basic functionality to communicate with agents asking for information on products requires the processing of incoming communication of asking agents and preparation of outgoing information. The incoming information may refer to scopes of interests of the asking agent, or to subscription. The communicated scopes of interest are identified using the knowledge base agent interest identification kb. Incoming communication on subscription is identified using knowledge base subscription identification kb. Outgoing communication containing product information to agents asking for information is prepared using knowledge base proposal communication info. All these knowledge bases are used within the component agent interaction management.

3. *Communication with agents providing information on products*

Communicated information on products can be processed in two different ways. First, the product information can be identified, using knowledge base provider info identification kb. Second, from the fact that information is provided on a product with certain characteristics, it can be abstracted (from the given product) that this provider is able to offer (at least some) products with these characteristics in general. This is done using knowledge base provider scope identification kb. Communication to an agent that may be able to provide information is prepared using knowledge base provider request kb. All these knowledge bases are used within component agent interaction management.

4. *Maintenance of acquired information on products*

The agent can identify that all communicated information on products has to be stored, using knowledge base provider info maintenance identification kb, within component agent interaction management. Moreover, by knowledge base observation info maintenance identification kb, within component world interaction management, new observation results on products to be stored can be identified.

5. *Maintenance of scopes of interest of agents*

The agent can identify that the incoming requests of agents are to be maintained. This functionality is specified by the knowledge base agent interest maintenance identification, used within component agent interaction management.

6. *Maintenance of scopes of products agents can provide*

Scopes of information agents can provide are stored, if the incoming communication is handled in an appropriate way using knowledge base provider scope maintenance identification kb, used within component agent interaction management.

7. *Own control*

Control of the agent's own processes is defined by the knowledge base focus kb, used within component own process control.

8. *Determining matches between products and scopes of interests*

To determine matches between products and scopes of interest the agent can use the knowledge bases attribute and scope kb, product scope abstraction kb, and strict match kb within component determine proposals.

Combinations of these functionalities define specific types of agents. For example, if a provider agent is designed, functionalities 2., 4., 5., 8. may be desired, whereas functionalities 1., 3., 6., 7. could be left out of consideration. If an agent is designed to support a user in finding information on products within a certain scope, functionalities 1., 3., 6., 8. (and perhaps 4.) may be desired, whereas 2. and 5. may be less relevant. For a mediating agent, or for an agent that has to play different roles, almost all functionalities (i.e., 2. to 8.) may be desired. The generic agent architecture introduced in Sections 3 and 4 can be instantiated in different manners to obtain, among others, the types of agents mentioned. The relation between the agent's basic functionalities, its knowledge, and where the knowledge is used is summarized in the table in Figure 2.

5.2 Reactive, Pro-active, and Other Forms of Behavior

Depending on the choices made, the broker can show reactive behaviour towards agents asking for information on products and provider agents.

- In reaction to an agent that asks for products within a certain scope, the broker determines which of the products it knows, fit to this scope. The available information on the resulting products is communicated to the agent (e.g., author, title, year, topics, abstract, location, reference).
- Once an agent interest is known to the broker agent, it is reactive to any information providing agent that announces a product that fits the agent's scope. In such a case the information on this product is communicated to this agent (i.e., to all relevant agents).

basic functionality	*knowledge specifying functionality*	*in component*
1. observation	observation initiative kb observation info identification kb	WIM WIM
2. communication with agents asking for information	agent interest identification kb subscription identification kb proposal communication kb	AIM AIM AIM
3. communication with agents providing information	provider info identification kb provider scope identification kb provider request kb	AIM AIM AIM
4. maintenance of product information	observation info maintenance identification kb provider info maintenance identification kb	WIM AIM
5. maintenance of scopes of interest	agent interest maintenance identification kb	AIM
6. maintenance of scopes of products agents can provide	provider scope maintenance identification kb provider scope abstraction kb attribute and scope kb	AIM MAI MAI
7. own control	focus kb	OPC
8. match between products and scopes of interests	attribute and scope kb product scope abstraction kb strict match kb	DP DP DP

Figure 2. Relation between basic functionalities and knowledge required.

Pro-active behavior occurs when the broker agent has as a characteristic that it is pro-active with respect to certain agents. A pro-active broker, from time to time takes the initiative to ask provider agents for information on products which match some of its subscribed request profiles. It may focus on an agent's scopes of interest and actively select information providing agents and ask them whether they have products that fit in one of these scopes.

The behavior of the broker agent may depend on other characteristics of the broker agent as well. In the example, the knowledge used within its own process control was kept rather simple. It is not difficult to extend this knowledge in such a way that more complex forms of pro-active social behavior are initiated and controlled. For example, it is also possible that the broker pro-actively determines an expected scope of interest of an agent and proposes products that fit this expected scope of interest.

6 Maintenance by Communication

The requirements on functionality of a broker agent may evolve over time; a user may first only ask for an agent that is able to communicate information on its own products, and later ask for more functionality. For example, functionality that allows the agent to look for information on products itself, and store this information. In the compositional agent model presented here, a substantial part of the agent's functionality is represented in a declarative manner by explicit knowledge. An important advantage of this form of representation is that this functionality specification can be adapted easily by replacing (parts of) the knowledge bases by others, and, even more important, the knowledge can easily be made a subject of communication. This can be exploited to design an agent that supports its own maintenance on the basis of (new or updated) knowledge that is communicated to it.

6.1 Communication of Maintenance Knowledge

The agent model has been extended with possibilities to support its own maintenance in the following manner. First, communication on knowledge has been added to the model by extending the information types dealing with communication by explicit representation of knowledge elements as terms:

knowledge_element(<kb-name>, <antecedent>, <consequent>)

Both antecedents and consequents are taken from the sort CONJ, which stands for conjunctions of literals. Besides knowledge it may be important to communicate which functionality is supported by a knowledge base. This is expressed by a relation

supports_functionality: kb *component* functionality

that can be used to represent the knowledge depicted in the table of Figure 2 in Section 5. Within the component agent interaction management a knowledge base is added to extract the maintenance information from the communication information:

Knowledge base maintenance knowledge identification kb

if communicated_by(knowledge_element(k:kb, a:conj, c:conj), pos, maintenance_agent)
then new_maintenance_info(knowledge_element(k:kb, a:conj, c:conj), pos)

if communicated_by(supports_functionality(k:kb , c:component, f:functionality), pos, maintenance_agent)
then new_maintenance_info(supports_functionality(k:kb , c:component, f:functionality), pos)

The extracted maintenance information is transferred to the agent component own process control and stored in this component using the relation maintenance info. It is also possible to model outgoing communication on maintenance. For example, if the broker agent decides that is needs a new functionality, this can be communicated to the maintenance agent:

to_be_communicated_to(required_functionality(f:functionality), pos, maintenance_agent)

The maintenance agent can react on this communication by communicating maintenance information as described above.

6.2 Controlling Maintenance in Own Process Control

Within the agent component own process control the (sub-)component own maintenance management is distinguished. Within this component the maintenance information received is analysed, and if found adequate, the decision is made to execute maintenance. The following knowledge is used:

if maintenance_info(supports_functionality(k:kb , c:component, f:functionality), pos)
and required_functionality(f:functionality)
then kb_to_be_added_in(k:kb , c:component)

if kb_to_be_added_in(k:kb , c:component)
and maintenance_info(knowledge_element(k:kb, a:conj, c:conj), pos)

then kb_elt_to_be_added_in(knowledge_element(k:kb, a:conj, c:conj), c:component)

Conclusions drawn using this knowledge are transferred to the component to which the reference is made, which actually executes the maintenance. Similarly, retraction of knowledge (if a functionality is not desired anymore) is modelled. After these changes the agent will show the functionalities related to the modified knowledge.

7 Discussion

Applications of broker agents (addressed in, e.g., Chavez *et al.* 1996, Chavez *et al.* 1997, Kuokka and Harada 1995, Martin *et al.* 1997, Sandholm and Lesser 1995, and Tsvetovatyy and Gini 1996), often are not implemented in a principled manner: without an explicit design at a conceptual level, and without support maintenance. The broker agent architecture introduced here was designed and implemented in a principled manner, using the compositional development method for multi-agent systems DESIRE (Brazier *et al.* 1997). An application to the design of an intelligent Website in insurance can be found in (Jonker *et al.* 2001). Due to its compositional structure it supports reuse and maintenance; a flexible, easily adaptable architecture results. Moreover, within the agent model facilities have been integrated that provide automated support of the agent's own maintenance. Therefore, the agent is not only easily adaptable, but, because adaptation is automated, it shows adaptive behaviour to meet new requirements (either in reaction to communication with a maintenance agent, or fully autonomous).

The approach to automated maintenance introduced here has a high potential. In future research this potential will be explored further. In current research the broker agent model is applied within a project on electronic commerce, in co-operation with the Internet application company Crisp, and in a project on intelligent Web-sites in co-operation with the insurance company Delta Lloyd and the software company Ordina Utopics. The approach introduced here addresses dynamic modification of the agent's own knowledge. Dynamic modification of the agent's own process composition (e.g., adding/ deleting components or information links) is another possibility, which may result in complete re-design of the agent; this is addressed in see (Brazier, Jonker *et al.* 1998).

Required properties or functionalities of agents can be formalised, and the relation between required properties and underlying assumptions can be established in a formal manner. An example of a result of such a formal analysis is the relation between basic functionalities (required properties) and available knowledge (assumptions) discussed in Section 5 (see Figure 2). In this paper the result of formal analysis was used in the agent model; the formal analysis itself was done by us as designers. To support this, a compositional verification method for multi-agent systems has been developed and successfully applied to verify the behaviour of a multi-agent system for one-to-many negotiation (see Brazier, Cornelissen *et al.*1998) and to give a formal analysis of pro-activeness and reactiveness (see Jonker and Treur 2001). One of the more ambitious aims of our future research is to explore possibilities to include these formal analyses themselves in an agent model, and not only the results obtained by them.

Acknowledgments

Frances Brazier, Mehrzad Kharami and Pascal van Eck made useful comments on an earlier draft of the text. Mehrzad and Pascal also supported experiments with some variants of the broker agent model.

References

Brazier, F.M.T., Dunin-Keplicz, B. Jinnings, N.R. and Treur, J. (1997), "Formal specification of Multi-Agent Systems: a real-world case," in V. Lesser (ed.), *Proceedings of the First International Conference on Multi-Agent Systems, ICMAS'95,* MIT Press, Cambridge, MA, 1995, pp. 25-32. Extended version in: *International Journal of Cooperative Information Systems,* M. Huhns, M. Singh, (eds.), special issue on Formal Methods in Cooperative Information Systems: Multi-Agent Systems, vol. 6, pp. 67-94.

Brazier, F.M.T., Cornelissen, F., Gustavsson, R., Jonker, C.M., Lindeberg, O., Polak, B. and Treur, J. (1998), "Compositional design and verification of a multi-agent system for one-to-many negotiation," in *Proceedings of the Third International Conference on Multi-Agent Systems, ICMAS'98*, IEEE Computer Society Press, pp. 622-629.

Brazier, F.M.T., Jonker, C.M. and Treur, J. (2000), "Compositional design and reuse of a generic agent model," *Applied Artificial Intelligence Journal,* vol. 14, pp. 491-538.

Brazier, F.M.T., Jonker, C.M., Treur, J., and Wijngaards, N.J.E. (1998), "Compositional design of a generic design agent," *Design studies journal*, 2001, in press.

Chavez, A., Maes, P., Kasbah (1996), "An agent marketplace for buying and selling goods," in *Proceedings of the First International Conference on the Practical Application of Intelligent Agents and Multi-Agent Technology, PAAM'96*, The Practical Application Company Ltd, Blackpool, pp. 75-90.

Chavez, A., Dreilinger, D., Gutman, R., and Maes, P. (1997), "A real-life experiment in creating an agent market place," in: *Proceedings of the Second International Conference on the Practical Application of Intelligent Agents and Multi-Agent Technology, PAAM'97*, The Practical Application Company Ltd, Blackpool, pp. 159-178.

Jonker, C.M., Lam, R.A., and Treur, J. (1999), "A multi-agent architecture for an intelligent website in insurance," in M. Klusch, O. Shehory, and G. Weiss (eds.), *Cooperative Information Agents III, Proceedings of the Third International Workshop on Cooperative Information Agents, CIA'99.* Lecture Notes in Artificial Intelligence, vol. 1652. Springer Verlag, pp. 86-100. Extended version in: *Journal of Applied Intelligence*, 2001, in press.

Jonker, C.M. and Treur, J. (2001), Compositional verification of multi-agent systems: a formal analysis of pro-activeness and reactiveness. In: W.P. de Roever, H. Langmaack, A. Pnueli (eds.), Proceedings of the International Workshop on Compositionality, COMPOS'97, Springer Verlag, 1998. Extended version in:

International Journal of Cooperative Information Systems, 2001, in press.

Kuokka, D. and Harada, L. (1995), "On Using KQML for Matchmaking," in V. Lesser (ed.), *Proceedings of the First International Conference on Multi-Agent Systems, ICMAS'95*, MIT Press, Cambridge, MA, pp. 239-245.

Martin, D., Moran, D., Oohama, H., Cheyer, A. (1997), "Information brokering in an agent architecture," in *Proceedings of the Second International Conference on the Practical Application of Intelligent Agents and Multi-Agent Technology, PAAM'97*, The Practical Application Company Ltd, Blackpool, pp. 467-486.

Sandholm, T. and Lesser, V. (1995), "Issues in Automated Negotiation and Electronic Commerce: Extending the Contract Network," in V. Lesser (ed.), *Proceedings of the First International Conference on Multi-Agent Systems, ICMAS'95*, MIT Press, Cambridge, pp. 328-335.

Tsvetovatyy, M. and Gini, M. (1996), "Toward a Virtual Marketplace: Architectures and Strategies," in *Proceedings of the First International Conference on the Practical Application of Intelligent Agents and Multi-Agent Technology, PAAM'96*, The Practical Application Company Ltd, Blackpool, pp. 597-613.

Wooldridge, M. and Jennings, N.R. (1995a), Agent theories, architectures, and languages: a survey. In: Wooldridge and Jennings 1995b, pp. 1-39.

Wooldridge, M. and Jennings, N.R. (eds.) (1995b), Intelligent Agents, *Proc. of the First International Workshop on Agent Theories, Architectures and Languages, ATAL'94 (Lecture Notes in AI, vol. 890)*, Springer Verlag.

Chapter 7

Collective Behavior Evolution in a Group of Cooperating Agents

J. Liu and J. Wu

This work addresses the issue of how to acquire a collective goal-directed task-driven behavior in a group of distributed agents. The specific problem that we consider here is how simulated ants can perform coordinated movements to collectively transport an object toward a desired goal location. We propose an evolutionary computation mechanism in which no centralized modeling and control are involved except a high-level criterion for measuring the quality of collective task performance. The evolutionary learning approach is based on a fittest-preserved genetic algorithm. While giving the formulation and algorithm of collective behavior learning, we also present the results of several computer simulations for illustrating and validating the effectiveness of the proposed approach.

1 Introduction

The motivation behind our present work on collective group task-handling stems from the fact that a distributed multi-agent approach has certain advantages over a single-agent approach, in that a multi-agent system can exhibit the characteristics of structural flexibility, reliability through redundancy, simple hardware, adaptability, reconfigurability, and maintainability.

In this chapter, we will describe an approach to generating collective ant behavior in performing an object-moving task. The difficulty of this task lies in that the movements of the ants with respect to an object cannot follow an explicit, global plan and control command, due to certain modeling limitations as well as planning costs. In such a case, it is important that the individual ants locally modify their movement strategies, and at the same time, create a desired collective interaction between the ant group and the object, which will successfully transport the object to a desired goal location. The proposed approach begins with a model of local interactions between the ants and the object in the environment, and then apply a genetic algorithm, as a global optimization technique, to select the local movement strategies of the individual ants, in an attempt to maximize the overall effectiveness of moving the object toward the goal location.

1.1 Related Work

1.1.1 Ant Systems

As inspired by ethological research on the behavioral characteristics of real ants, Dorigo *et al.* first proposed a computational model of cooperating agents, called "ant system" (Dorigo *et al.* 1996). In this system, each ant selects a path from its colony to the feeding source by following the accumulated quantities of pheromone as laid by other passing ants. Finding paths in such a cooperative way can effectively solve classical combinatorial optimization problems, as has been demonstrated and reported in (Maniezzo *et al.* 1994, Dorigo and Gambardella 1996, Dorigo and Gambardella 1997).

A more complete account on how to design and implement swarm-based problem solving systems as well as a comprehensive review on recent research progress can be found in (Bonabeau *et al.* 1999).

Among others, White *et al.* extended the ant system by incorporating a self-adaptation algorithm in generating and selecting the ant controlling parameters (White *et al.* 1998).

In our present work, we are also concerned with the computational model of ant-like systems. Our focus is on how to evolve desirable goal-directed collective behavior among cooperating ants.

1.1.2 Collective Behavior Learning

In multi-agent systems research, one of the important issues is to enable individual agents to dynamically acquire a collective task-handling behavior that adapts to the changes in their task environment.

While the methodology for behavior engineering of a single agent has been established, for instance, by following the stages of target behavior analysis, architecture design, implementation, training, and behavior assessment (Colombetti *et al.* 1996), the learning of complex group behavior largely remains to be an open problem.

Fukuda and Iritani (Fukuda and Iritani 1995) proposed a mechanism for modeling group cooperative behavior among decentralized autonomous agents, called CEBOT (i.e., Cellular Robots). Their work simulated the generation of group behavior based on a globally stable attractor and the identification of new group behavior based on bifurcation-generated new attractors.

Mataric (Mataric and Cliff 1996, Mataric 1997) studied the problem of group behavior such as coordination among agents, and developed a group behavior learning method in which heterogeneous reward function-based reinforcement learning (RL) was applied to synthesize collective behavior, such as flocking, foraging, and docking by means of direct/temporal summation and switching of some basic behavioral patterns. The RL technique is useful when dealing with problems of limited search space, e.g., a few possible combinations of individual agent movements. If an application involves a significantly large search space, RL-based learning will become quite time-consuming. This is primarily due to the fact that all external stimuli encountered by agents must be distinctly identified and that their corresponding responses must be sufficiently conditioned or reinforced.

In his recent dissertation, Balch (Balch 1998) studied a methodology for building and evaluating learning in behavior-based robot teams. In his work, he showed that the quantitative impact of different reinforcement functions on learning different behaviors in performing agent foraging, soccer, and cooperative movements.

Cao *et al.* (Cao *et al.* 1997, Cao *et al.* 1995) identified five major themes of agent group studies in task domains such as traffic control, object-moving, and foraging. The five themes are group control architecture (e.g., decentralization and differentiation), resource conflict resolution (e.g., space sharing), the origin of coordination (e.g., genetically determined social behavior *vs.* interaction-based coordinated movement (Mataric 1994)), learning (e.g., control parameter tuning for desired coordination), and geometric problem solving (e.g., geometric pattern formation).

In the case of learning coordinated movements by distributed ants, the possible combinations of ant local movements can be enormous. Thus, in our study, as an alternative to RL-based learning, we will use a GA-based approach. This is because GA, as a global optimization technique, is effective in handling a large number of diversified (sometimes redundant) movement patterns as created by the ants, by means of evolutionary selection.

1.1.3 Application of Genetic Algorithms in Robotics

A genetic algorithm (GA) (Goldberg 89, Holland 75) is based on a computational model of natural evolutionary processes, which may be applied to solve computation-intensive problems, such as numerical search and optimization (Back *et al.* 1997, Schwefel 1995). Robotics is one of the fields in which researchers have found many applications of GA, ranging from robot morphology generation, control strategy synthesis, to geometric motion planning (Fujji *et al.* 1996, Harvey *et al.* 1996, Husbands *et al.* 1996, Mataric and Cliff 1996, Husbands *et al.* 1995, Miglino *et al.* 1997, Nolfi and Parisi 1996). In what follows, we will take a look at some related work in evolutionary robotics.

Evolving Agent Controllers

Beer and Gallagher (Beer and Gallagher 1992) explored the idea of evolving continuous-time recurrent neural networks in order to achieve adaptive agent control; an example was given in the case of locomotion-control for a six-legged insect-like agent. Yamauchi and Beer (Yamauchi and Beer 1994) also studied the evolution of neural network controllers for autonomous agents, coupled with sequential and learning behavior. Colombetti and Dorigo (Colombetti and Dorigo 1993) developed a classifier system (CS) based approach to agent control that relied on the construction of a hierarchical architecture of CS. Miglino, Lund, and Nolfi

(Miglino et al. 1997) investigated the validity of evolving neural controllers based on computer simulation and transferring them to a Khepera robot.

Unlike our present work, the above-mentioned studies focused on the automatic construction or programming of a single-agent controller, whereas we emphasize on learning in a group of distributed agents for achieving a goal-directed collective behavior.

Genetic Programming

Genetic programming focuses on the synthesis of agent motion/control programs by means of a genetic algorithm. Koza (Koza 1992) showed the application of genetic programming in constructing a subsumption architecture for agents capable of performing wall-following and object-moving tasks. Reynolds (Reynolds 1994) also used genetic programming to create control programs that enabled a mobile vehicle to avoid collisions.

In our present work, we are concerned with evolving collective behavior in a group of individual ants with respect to some global goal. This problem may be viewed as one of the genetic programming problems aimed to synthesize local movement strategies of the individual ants from their previous spatial configurations. The movement strategies are continuously selected in order to optimize certain global performance (fitness) of the group in a collective task.

1.2 The Organization of the Chapter

The chapter is structured as follows: Section 2 provides a formal statement of the problem addressed in our work. Section 3 gives an overview of the proposed approach to group behavior learning with both modeling and algorithmic details. Sections 4 and 5 describe two sets of computer simulations. The first illustrates how the group behavioral learning works, with point ants and object and with a simplified dynamics model of ant-object interaction. The second presents some physically-modeled, collective object-moving examples to further investigate the effectiveness of learning in a more complex pushing task, where ants of rectangular shape can have many possible point contacts while pushing and orienting an object. Section 6 concludes the chapter by highlighting the main contributions of the present work.

2 Problem Statement

The studies mentioned in the preceding section have, to some extent, shared one thing in common; namely, the motion of the ants is planned and executed by utilizing complete and/or partial representations of ant environment such as object shape, obstacle dimensions, and the position and orientation information about other ants. The question that remains is how to effectively generate the coordinated motions for a group of ants when the information about the ant environment, such as the states of other ants with respect to their local environment conditions, is not available or too costly to obtain. This issue is particularly relevant if we are to develop autonomous agents that can work collectively and adaptively on a shared common task, even though each agent can only sense and hence react to its environment locally.

2.1 What is the Collective Behavior of an Ant System?

Let $\varepsilon(t)$ denote the current environment state of an ant team at timestep t, ξ denote the sensory inputs of the individual ants. In the environment, each ant is capable of performing some actions τ. Coordinated movements, Ω, by the ants in response to $\varepsilon(t)$ is composed of actions τ such that they satisfy μ, the constraints imposed by the environment such as point contacts with an object, and ν, the criterion of the collective task performance. This may be expressed as follows:

$$\forall\varepsilon(t),\ \exists\,\gamma(t) \in \tau,\ \ s.t.\ \Omega \equiv \{\,\gamma(t)\,\},\ \ E(A) \stackrel{\Omega}{\Longrightarrow} \nu \qquad (1)$$

where $\gamma(t)$ corresponds to a goal-directed group behavior at timestep t. A denotes individual ants. $E(A)$ denotes the performance evaluation of A.

2.2 What is Group Behavior Learning?

The problem of group behavior learning is essentially a problem of learning to collectively perform a given task. Solving this problem requires ants to continuously acquire their experience for their local interaction with the environment. Such group experience is incrementally updated. This may be expressed as follows:

$$\epsilon(t) \longleftarrow \varepsilon(t-1) \oplus \epsilon(t-1); \quad \Omega \equiv \epsilon(t) \mid_{\Delta\epsilon(t) \approx 0,\, E(A)=\nu} \qquad (2)$$

where $\epsilon(t)$ represents the group experience at timestep t. $\oplus$ denotes a computational updating mechanism. $\varepsilon(t-1)$ correspond to the environment state sensed at timestep $t-1$.

An ant group behavior will be generated from the group experience following a series of environment state transitions. Therefore, the process of group behavior learning may be regarded as the process of converging ant experience.

The specific group learning problem to be solved in this work may be stated as follows: Given a group of three individual ants, A, capable of sensing and changing their own local positions and orientations, $\varepsilon(t)$, with respect to an object and a desired goal location in their workspace, how to develop a genetic algorithm-based mechanism, $\oplus$, such that A can gradually acquire coordinated local movements, Ω, based on a series of $\epsilon(t)$ changes. It is required that as a result of Ω, the object will be collectively pushed by A toward a goal location, $\varepsilon(t)'$.

As in some real-world situations, in our present study we assume that each ant has limited sensing, communication, and on-board computation capabilities. More specifically,

1. **Sensing** The group ants are capable of distinguishing which is the object to be pushed and which is its group mate (e.g., through color detection). Due to the limitation in its sensing range, the ant will not be able to detect the exact shape of the object. The ant can only sense the direction of the common goal location – hence estimate its relative orientation – but not the exact distance from the goal.
2. **Communication** There will be no cross communication between ants.

3. **On-board computation** The on-board computation of an ant is responsible only for its motor-level control, local sensing, and agent communication.

Besides the above assumptions, it should be pointed out that in order for us to focus on the issue of learning collective object-moving behavior, we assume that each ant is capable of finding the object whenever the contact between the ant and the object is lost. With respect to a collective object-moving task, we do not consider the problem of finding the optimal number of individual ants for coordination.

3 The Proposed Approach

This section presents a genetic algorithm-based approach to the problem of learning how to perform a collective object-moving task by a group of distributed ants.

3.1 The Basic Ideas

In our present work, we apply a fittest-preserved genetic algorithm to implement group behavior learning. The fitness function of our genetic algorithm involves two components: one corresponds to the local interactions, i.e., the relative spatial configurations of individual ants relative to an object, whereas the other corresponds to the global feedback information, i.e., the direction of a net pushing force on the object relative to a goal direction, as a result of those ant spatial configurations. Based on the current spatial configuration of the ants, the genetic algorithm will first generate a set of new spatial configurations, evaluates these configurations using the fitness function (as denoted by a dashed line), and then selects the high-fitness configurations for the next step of evolution. In doing so, the local movement strategies for the individual ants are being continuously evolved and performed. Each strategy is encoded in terms of the local relative position and orientation of a group.

3.2 The Ants

In our work, simulated ants will be used, which are graphically represented as circular or rectangular shapes[1]. Each ant is equipped with an array of sensors to sense an object of a certain height in its workspace environment. During collective object-moving, the information about a global goal for the object will be known to all individual ants in order for the ants to estimate their relative positions and orientations. Apart from the desired goal location, another message will also be continuously communicated, that is, the local movement strategies for the ants in the next step.

3.3 Performance Criterion for Collective Object-Moving

The task performance of ants in collective object-moving will be evaluated based on a high-level criterion. This criterion specifies the following collective performance requirements:

1. The ants should maximize their net pushing force on the object.
2. The direction of the object movement should be pointed toward a desired goal.
3. Whenever a goal location is redefined, the ants should change their movement strategies accordingly.

The above-mentioned criterion for the task is to be met by ants based on their coordinated movements.

3.4 Evolving a Collective Object-Moving Behavior

In order to enable the learning of high-level collective intelligence for coordination among ants in their object-moving, we utilize an evolutionary computation mechanism that evaluates possible local motion strategies, and then broadcasts to the ants the highest-fitness ones according to the criterion of collective task performance. Based on such a mechanism,

[1]Here we consider rectangle-shaped ants in order to make the object-moving problem more complex and interesting. In the rectangular case, the configurations of the contacts between an ant and an object being pushed will not be unique.

although the ants remain to behave locally, they can produce a certain amount of collective intelligence, for instance, creating a greater force in order to push the object, and adjusting their net force in order to create a smooth trajectory for the object.

The evolutionary computation mechanism for selecting local ant movement strategies is based on a fittest-preserved genetic algorithm. This algorithm performs a stochastic search based on a model of fitness-based natural evolution. The feature of the algorithm consists in that it emphasizes on the highest-fitness individual[2], i.e., the best movement strategy, by recording it directly to the next generation and applying mutation to this individual without any crossover operation. The complete algorithm as used in the group behavioral learning is presented in Figure 1. The explicit semantics of variables in the algorithm will be given in the following section.

4 Collective Object-Moving by Applying Repulsive Forces

In order to illustrate the key steps of the above-mentioned group behavioral learning approach and to show its general features, this section presents the formulation and experimentation in an illustrative collective object-moving example. In this example, we assume that the ants can be regarded as point ants. In addition, we define the dynamics of ant-object interaction by introducing a notion of artificial repulsive force between an ant and an object to be pushed.

4.1 A Model of Artificial Repulsive Forces

The model of artificial repulsive forces between ants and an object is based on a Hooke's-law spring-like model as follows: If an ant moves inside an l_0-radius region around the object, it will immediately exert an artificial repulsive force $\vec{F_i}$ on the object. The direction of $\vec{F_i}$ is shown in

[2] A distinction should be made between individuals in the populations and the individual ants.

```
begin
  define a fitness function (S_f),
  define the maximum number of generations to be evolved (G),
  define a population size for selection (P),
  define the probability of crossover operation (p_c),
  define the probability of mutation operation (p_m),
  create an initial population of P members.
  for generation : 1 ⟶ G do
    evaluate fitness S_f in the current generation:
      for population : 1 ⟶ P do
        propose new spatial configurations for ants,
        compute the corresponding motions of the ants,
        evaluate the fitness values for the computed movements.
      endfor
    record the highest-fitness individual,
    mutate the highest-fitness individual with p_m to get a
      new individual for the next generation,
    select additional P − 1 individuals based on their fitness values,
    apply a two-point crossover operation to the P − 1 individuals
      to get P − 1 new members,
    mutate the P − 1 new members with probability p_m
      to create the next generation.
  endfor
end
```

Figure 1. The genetic algorithm that is used to compute, evaluate, and select collective goal-achieving, object-moving movement strategies for ants, by selecting, recombining, and mutating a population of potential movement strategies.

Figure 2, and the scale of $\vec{F_i}$ is computed as follows:

$$F_i = \begin{cases} \eta \times (l_0 - d_i) & d_i \leq l_0, \\ 0 & otherwise. \end{cases} \tag{3}$$

where η is a positive coefficient, and d_i is the Euclidean distance between the ant and the object at the time of exerting a repulsive pushing force. If the ant is outside the l_0 range of the object, the pushing force will become zero.

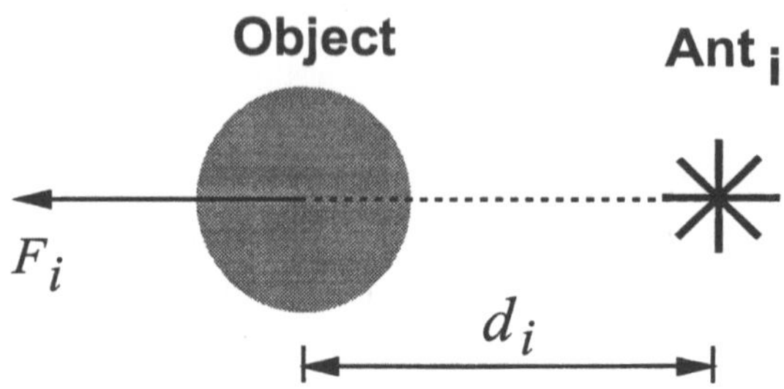

Figure 2. A top view of collective object-moving (Note: for the sake of clarity, only one ant is depicted) where ants, marked with '*', move toward an object, signified by a grey round area. When their distances d_i are less than or equal to l_0, *a threshold*, pushing forces will be exerted on the object by the ants, according to the spring-like model of repulsive pushing force (refer to the text for details).

By introducing a notion of artificial repulsive forces, we will regard the interaction between three ants and an object as being created by such forces. In other words, the ants will collectively "push" the object through their repulsive forces, if and only if their distances to the object are less than or equal to a predefined constant, l_0.

4.2 Moving Force and the Resulting Motion of an Object

In this work, we are interested in the coordination among three ants in moving an object toward a desired goal location. The pushing force acting on the object will be contributed by all three ants. Here, we will apply the rule of vector composition to combine the pushing forces as created by the ants.

We will consider that the motion of an object is caused by the collective net repulsive pushing force on the object, $\sum \vec{F_i}$, as follows: The direction of object displacement is the same as the direction of the net force, and the magnitude of object displacement is proportional to that of the net force, as calculated by the following expression:

$$\mathcal{F} = \alpha \times \|\sum_i \vec{F_i}\| \tag{4}$$

where α is a positive coefficient and $\| \bullet \|$ denotes the modulus.

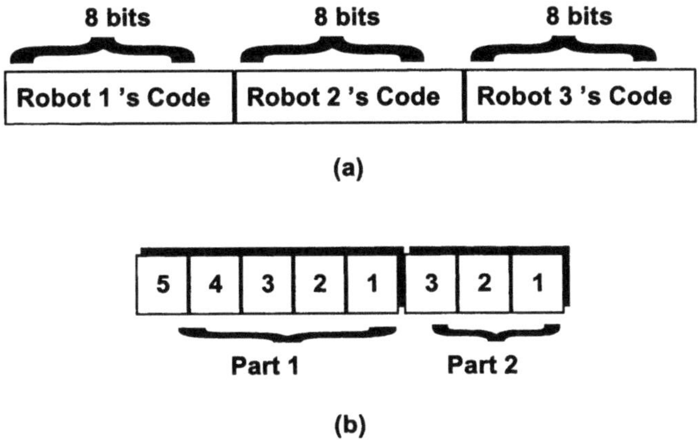

Figure 3. (a) A 24-bit binary string is used to represent individual members, i.e., chromosomes, in the genetic algorithm-based movement strategy selection for three ants. (b) Each of the 8-bit substring in the chromosome, encoding the local movement strategy of a single ant, consists of two parts: part 1 encodes the relative orientation angle β_i and part 2 local distance d_i.

4.3 Chromosome Representation

In the genetic algorithm as shown in Figure 1, each of the members in a population is encoded using a Gray code, called a chromosome. Specifically, we assign an 8-bit substring to encode the relative spatial configuration of one of the three ants, as shown in Figure 3(a). Furthermore, we subdivide the 8-bit substring into two parts, as in Figure 3(b), that correspond to orientation angle β_i and distance d_i of an ant relative to an object, respectively, as illustrated in Figure 4. The most significant 5 bits of the 8-bit substring are assigned to encode β_i. That is, the entire range of 2π for the relative orientation is divided into 32 units with an angular resolution of $\frac{\pi}{16}$. The rest 3 bits are assigned to encode d_i. Thus, the distance is divided into 8 units with a linear resolution of $\frac{l_0}{8}$.

Also, it may be noted from Figure 4 that β_i and d_i will together form a polar coordinate for the ant. In this polar coordinate system, the polar axis is originated at a certain reference point on the object, pointing toward the goal, as denoted by '+'. Thus, if ant i has distance d_i from the object and orientation β_i in this polar coordinate system, then the relative spatial

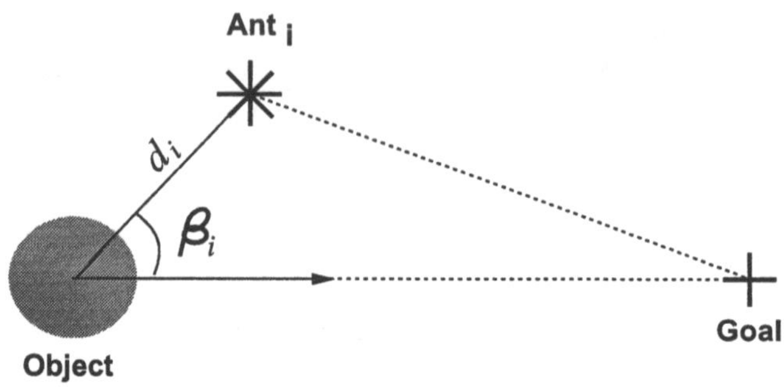

Figure 4. A top view of ant local spatial configuration in a movement strategy (Note: for the sake of clarity, only one ant, marked with '*', is shown). In our experiment, once an ant has moved to a location close enough to an object, i.e., distance to the object $d_i < l_0$, it can exert a repulsive force on the object at its relative orientation, β_i, with respect to a reference point on the object and a desired goal location. β_i and d_i can be regarded as the polar coordinates of the ant in a coordinate system defined by the current and the desired locations of the object.

configuration of ant i can be expressed as follows:

$$\vec{\mathcal{P}}_i = d_i \times e^{j\beta_i}. \tag{5}$$

Accordingly, the direction of a pushing force, $\vec{F}$, as produced by the ant will become $e^{-j\beta_i}$. The magnitude of this force will satisfy Eq. 3. Thus, we have:

$$\vec{F_i} = F_i \times e^{-j\beta_i}. \tag{6}$$

4.4 Fitness Function

In the genetic algorithm as used for evolving ant coordinated movement strategies, a function is used to continuously evaluate the fitness of each member, i.e., a potential strategy, in a population, which is referred to as a fitness function. This function represents a high-level criterion, as stated earlier, for the task of collective object-moving. Specifically, our fitness function is defined as follows:

$$S_f = s_1 \times s_2 \times s_3. \tag{7}$$

where s_1 specifies a net pushing force on an object as created by ants, which can be mathematically expressed as:

$$s_1 = \alpha \times \|\sum_i \vec{F_i}\|. \tag{8}$$

s_2 measures how close the ants are to the object, which is calculated as follows:

$$s_2 = \gamma \times [\sum_i d_i]^{-1}. \tag{9}$$

Finally, s_3 indicates whether or not and how much the motion of the object is directed toward a desired goal location, or more specifically:

$$s_3 = \begin{cases} cos\theta & \theta \in [-\frac{\pi}{2}, \frac{\pi}{2}], \\ 0 & otherwise. \end{cases} \tag{10}$$

where θ is the direction of the net force by the three ants.

4.5 Experiments with Simulated Ants

In this section, we will describe our experiments in which the early-mentioned GA-based group behavior learning mechanism is implemented and validated. In order to have a closer examination into the process of coordinated movement, we will present and trace several snapshots of the behavioral selection and motion execution in a typical case study. In the experiments, the ants are designed such that they will exert repulsive forces simultaneously.

4.5.1 Task Environment

In our experiments, we define the workspace for three ants as a square arena, $\mathcal{R}$, of size 10 l_0 $\times$ 10 l_0. Whenever the ants or the object hit the boundary of the workspace, the task of object-moving will be reset. At the beginning, the three ants and the object will be randomly placed inside $\mathcal{R}$. A desired goal location will be arbitrarily generated within the workspace.

4.5.2 Simulation Results

Figure 5 provides a series of snapshots showing the process of group behavior learning based on a genetic algorithm, from generations 4 to

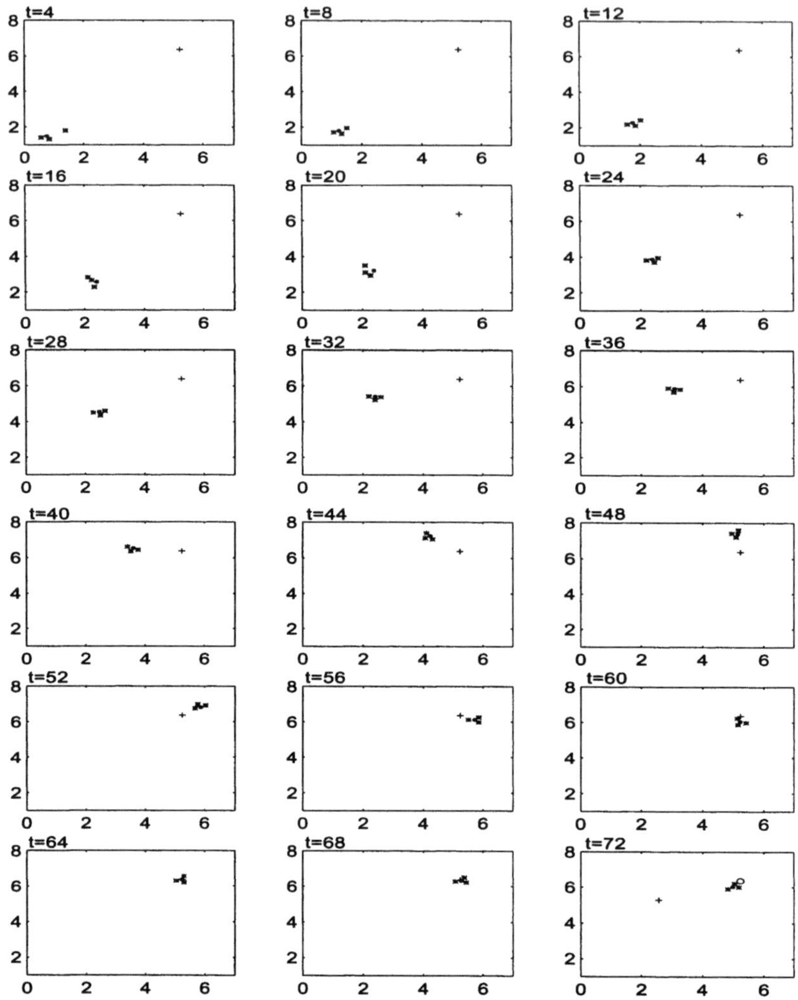

Figure 5. The snapshots of collective object-moving by three ants taken every 4 generations from generations 4 to 72 (top view). The collective movement in each snapshot is generated based on *the fittest member of a population* at a certain generation of collective behavior evolution. In the ant environment, a desired goal location is marked with symbol '+'. An object, marked with '•', is gradually moved toward the goal direction, after about 40 generations of movement strategy selection.

72 (taken every 4 generations). From these snapshots, we note that three ants continuously modified their spatial configurations in order to satisfy a high-level fitness requirement. As a result, an object, marked with '•', was gradually moved toward a desired goal location, marked with '+'. The object reached the goal location at generation 64.

4.5.3 Generation of a Collective Moving Behavior

Figure 6 presents the trajectories of three ants moving an object. In the figure, the solid line denotes the trajectory of the object, and the dashed line, dotted line and dash-dot line correspond to those of the three ants. We may note that at the beginning of object-moving, the ants moved in a rather chaotic manner. It took more than 20 generations for the group of distributed ants to emerge an effective group behavior satisfying the fitness function as given in Eq. 7. The motion trajectories were gradually getting smoother as the evolution of collective movement strategies went on. After about 40 generations, as marked in the figure, the coordination among the ants became most apparent. We also note that the group created a spiral motion near the goal location. This is because in many high-fitness individuals selected, i.e., effective strategies, the direction of a net repulsive force may not be directly pointing toward the goal location, implying that the GA-based learning after the given period of evolution yielded a good, but not necessary the best, solution for accomplishing the task.

4.5.4 Adaptation to New Goals

In our present case study, we are also interested in whether ants can adapt to the change of a goal location. Figure 7 presents the traced motion trajectories of the ants after the goal was changed to a new location at generation 70. In such a case, the ants immediately brought the object toward the new goal location, indicating that the GA-based evolutionary behavior learning can adapt to the changes in the task environment. In Figure 7, symbol 'o' at t=70 denotes a previous goal location.

The above observations are also reflected in a fitness function plot for the best individual of each generation over the entire process of behavior learning, as given in Figure 8. From the figure, we may identify four

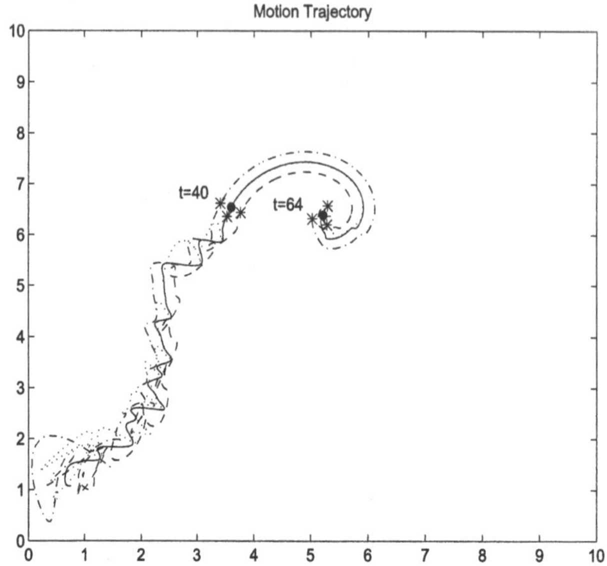

Figure 6. The object-moving trajectories created three ants. The solid line corresponds to the trajectory of the object, whereas the others correspond to the movement traces for the three ants. At the beginning, the net pushing force of the ants resulted in a rather randomized motion of the object. This corresponds to the random initiaization of potential collective movement strategies in the genetic algorithm. Next, the motion of the object was transformed into a phase of chaotic oscillation, as the genetic algorithm was undergoing the recombination and mutation of some sub-optimal local movement strategies (that is, certain terms in the fitness function may first get optimized before the others). Later after some generations of selection, a niche was found, representing a globally near-optimal collective movement strategy, and hence, the oscillating movement of the object became stabilized.

distinct stages in the evolution. The first stage occurred during the early 20 generations, where the ants executed their movement strategies quite chaotically. The reason is that the chromosome had just been randomly selected, and no high-fitness configuration was found with the genetic operations. The next stage happened in the following 19 generations, i.e., from generations 21 to 39. During this period, the ants executed several collective group behaviors of low-fitness. Then, during later task performance, from generations 40 to 64, improved collective behavior started to emerge, even though there were few slight oscillations involved. Finally, after generation 64, the fitness values reached a new level, even

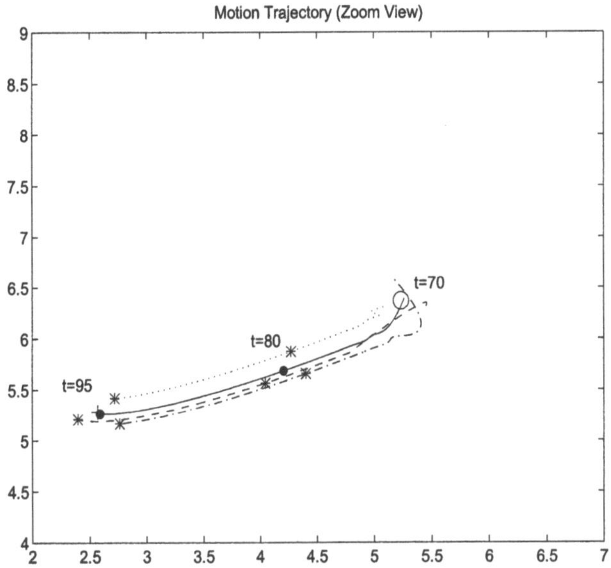

Figure 7. At generation 70, a new goal location was set. The earlier selected movement strategies for ants to collectively push an object can be immediately re-adjusted, since only the movement direction needed to be offset in this case.

when the goal location was changed twice, one at generation 70 and another at generation 97.

4.5.5 Discussions

In the formulation of our fitness function, we have emphasized on keeping a good balance between the local interaction of individual ants and the global goal-directedness of object movement. This is best reflected in the three terms of Eq. 7. Now let us consider what would happen if either the local interaction or the global goal-directedness were removed in the fitness function.

As we may recall, s_1 and s_2, as defined in Eq. 8 and Eq. 9, respectively, imply that all ants should try to surround the object as toughly as possible while maintaining a maximal pushing force. If only these two terms were involved in the task performance criterion, the ants would immediately acquire a series of pushing movements without taking into account any specific direction of pushing. In such a case, the object would be pushed

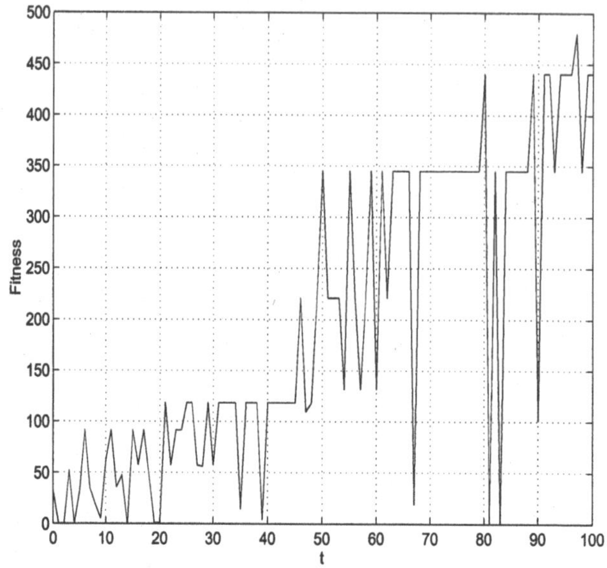

Figure 8. A fitness function plot for *the fittest member of a population* at each generation, over 100 generations, i.e., the entire evolution process of collective object-moving. Due to the fact that any slight movement change can cause significant variations in some of the terms in the fitness function, e.g., the direction and/or the magnitude of the net pushing force, the fitness values recorded have shown some fluctuations. Nevertheless, the trend of the fitness change tends to increase over the period of collective behavior evolution, even when the desired goal locations were reset twice, at generations 70 and 97.

straight toward the boundary of the ant workspace – any turning can be against previously evolved behavioral patterns and hence of low-fitness.

Unlike s_1 and s_2, s_3 in Eq. 7 implies that the net pushing force as created by ants should be pointed toward the direction of a desired goal location without concerning the exact distance from the goal. This term of global goal-directedness information alone will by no means create any effective pushing coordination among the ants. In such a case, even though the object may be brought to the goal location eventually, but the local motions of the ants will tend to become very chaotic.

From the above discussions, we note that there should be a good balance between a requirement on the overall goal-directedness and a requirement on the local movement behavior.

Another remark to be made is that by incorporating a constraint on the local movement behavior into the fitness function, we have embedded certain designer's knowledge in the fitness function. However, such high-level knowledge may not be complete as in many GA applications. This has also been reflected in the above-mentioned experiment that resulted in a near-optimal solution. In the case studies, we note that although the collective object-moving task was achievable, the fitness values appeared to be very fluctuating due to the definition of the three-term fitness function. From Eqs. 8 and 9, some slight changes in the relative position and orientation of the ants may produce quite dramatic changes in the group fitness value.

5 Collective Object-Moving by Exerting External Contact Forces and Torques

The aim of the previous example was to illustrate how the proposed group behavior learning approach works and what are the general characteristics of this approach. Towards this end, we simplified the dynamics model of ant-object interaction by introducing a notion of repulsive force and at the same time treated ants and an object as points of zero physical size.

In some earlier studies conducted by Mataric (Mataric et al. 1995), it was shown that where on an object an agent was pushing can, to a certain extent, affect the performance of a pushing task. In this section, we will consider the problem of group behavior learning to perform a complex object-moving task, where rectangular ants and a round/square object are involved that can have many possible point contacts among them. The goal of this example is to further examine and validate the effectiveness of learning in a carefully designed, more realistic simulation that is closer to a real-world situation than the preceding case study.

5.1 Interaction between Three Ants and an Object

Our next example involves three ants and one object. Two types of objects are considered: one is cylindrical (i.e., an object with a round-shaped cross-section, referred to as a round object) and another is cubic (i.e., an object with a square-shaped cross-section, referred to as a square object). The objects are rigid bodies, whose motions are based on the influence of forces and torques externally applied by the ants.

In order for ants to move an object, the ants must simultaneously exert external forces and torques on the object. The individual forces and torques arise from the point contacts between the ants and the object. Here we disallow collisions among the ants and the object. We assume that the ant always produces a unit force if it is in contact with the object in a certain direction. The net contact force on the object by the three ants can produce a linear displacement of the object. A contact force may be applied to the object, at some distance from its center of mass, which may create a corresponding torque. If the net torque on the object is non-zero, the object will rotate.

5.2 Case 1: Moving a Round Object

5.2.1 Moving Position and Direction

In the case of moving a round object, the contact positions and directions of ants can be described as follows: Assume that there are $\mathcal{N}$ positions around the object at which the ants may contact and push, and each ant can produce a pushing force in one of $\mathcal{M}$ directions. All these locations and directions are evenly distributed around the object and the ant. Thus, the resolutions for describing the pushing positions and the pushing-force directions are $\frac{2\pi}{\mathcal{N}}$ and $\frac{2\pi}{\mathcal{M}}$, respectively.

5.2.2 Moving Force and Torque

Figure 9 illustrates a contact pushing force on a round object that leads to its motion. Suppose that the direction of the pushing force is θ, and ant i pushes the object at a position with angular measurement β. In this case, the object will be pushed to move by unit force $\vec{F_i}$ and torque T_i, i.e.,

$$\vec{F_i} = \mathcal{F} \times e^{j\theta_i} \tag{11}$$

$$T_i = \mathcal{F} \times d_i \tag{12}$$

where $\mathcal{F}$ is the magnitude of the unit force, and d_i is the distance from the center of mass to the direction of pushing force $\vec{F_i}$.

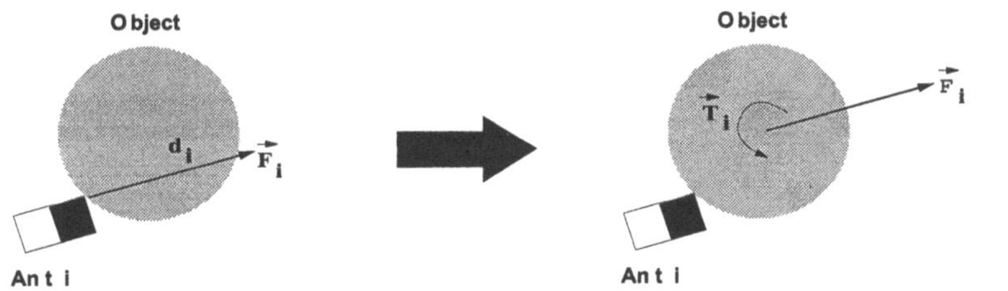

Figure 9. A top view of ants moving a round object (Note: for the sake of clarity, only one of the three ants is shown). The pushing force on the object is created by individual, rectangular-shaped ants via their contact points. As a result of this force, the object may be pushed to translate as well as rotate about a certain axis.

With the pushing force and torque mentioned above, the object will be pushed to move in the direction of the force, and at the same time, rotate about its center. The linear and angular displacements are defined as follows:

$$\int = \alpha_1 \times \mathbf{F} \tag{13}$$

$$R_\alpha = \alpha_2 \times \mathbf{T} \tag{14}$$

where α_1 and α_2 are positive coefficients, and **F** and **T** are joint force and joint torque, respectively.

5.3 Case 2: Moving a Square Object

As in the preceding case, ants will push an object through their contact forces. However, the present case differs from the above in that a square object is used, whose orientation will no longer be unique.

5.3.1 The Coordinate System

Figure 10 presents a top view of ants moving a square object. For the sake of clarity, only one of the three ants is shown. During collective object-moving, the relative spatial configuration of each ant is specified by a relative orientation, β_i, with respect to a desired goal location and some reference point on the object. The coordinate system for describing β_i is given in Figure 10. In this system, the origin lies in the reference point on the object, and the axis points toward the desired goal.

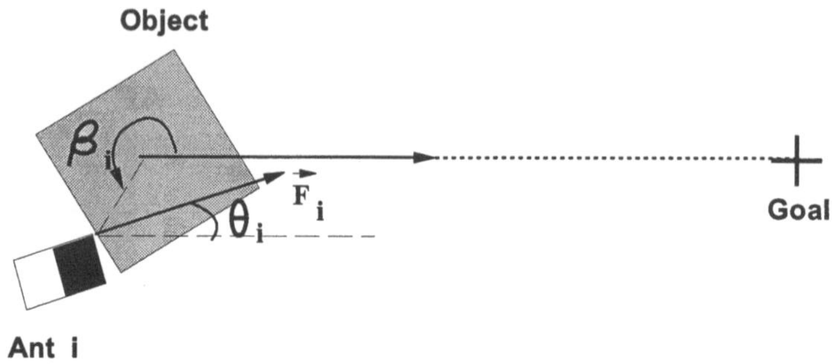

Figure 10. A top view of ants moving a square object (Note: for the sake of clarity, only one of the three ants is shown). As in the preceding case, the pushing force on the object will be created via the physical contacts between the ants and the object. During collective object-moving, the relative spatial configuration of each ant is specified by a relative orientation, β_i, with respect to a desired goal location and some reference point on the object.

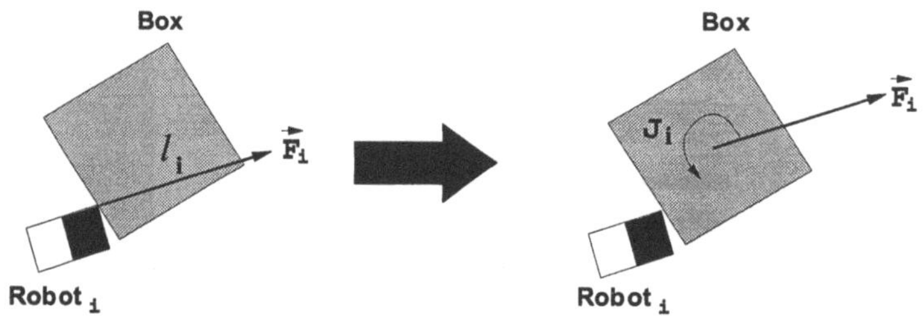

Figure 11. As a result of the net contact force on an object, as created by three ants, the object may be pushed to translate as well as rotate about a certain axis (Note: for the sake of clarity, only one of the three ants is shown).

5.3.2 Moving Force and Torque

Figure 11 shows a contact-pushing force on a square object, which causes the motion of the object. In this case, the definitions of force, torque, and displacements are the same as those in the preceding round-object case.

5.4 Chromosome Representation

In the present case studies, the chromosome of each individual ant is encoded in 10 bits. Thus, each member of a population, expressing a possible collective movement strategy for three ants, is represented using a 30-bit chromosome. In each 10-bit substring, the most significant 5 bits express angle β_i for one of the three ants in a polar coordinate system, as mentioned above. That is to say, the total range 2π of orientation is subdivided into 32 units with an angular resolution of $\frac{\pi}{16}$. The remaining 5 bits encode the direction of pushing force acting on the object by that ant. In this work, the ant can select a certain contact pushing direction from a total of 32 directions.

5.5 Fitness Functions

In the genetic algorithm for evolving the collective movement strategies of ants, a function is used to continuously evaluate the fitness of each member in a population. This function expresses a high-level criterion, as stated earlier, for the task of collective object-moving.

Case 1: Moving a Round Object

In this case, we measure an optimal group behavior using a fitness function of ants as follows:

$$S_f = s_1 \times s_2 \times s_3 \tag{15}$$

where

$$s_1 = \|\sum_i \vec{\mathbf{F}}_i\| \tag{16}$$

$$s_2 = 1.01 + cos\Theta \tag{17}$$

where

$$\sum_i \vec{\mathbf{F}}_i = \|\sum_i \vec{\mathbf{F}}_i\| \times e^{j\Theta} \tag{18}$$

$$s_3 = \|\sum_i \vec{\mathbf{T}}_i\|^{-1} \tag{19}$$

where Θ is the direction of the net contact force by all three ants. $\vec{\mathbf{F}}_i$ and $\vec{\mathbf{T}}_i$ denote a contact force and a corresponding torque created by ant i, respectively.

In the above definition of the fitness function, s_1 implies that the object should be pushed by a large net contact force, s_2 implies that the object should be moved toward a desired goal location, and s_3 implies that the rotation of the object should be discouraged during collective pushing.

Case 2: Moving a Square Object

In the case of moving a square object, we calculate a fitness function for ants as follows:

$$S_f = \sqrt{s_1} \times (s_2 \times s_3)^4 \tag{20}$$

where s_1, s_2 and s_3 are given in Eqs. 16, 17, and 19, respectively.

5.6 Experiments with Simulated Ants

This section presents some experiments in which the above-mentioned GA-based group behavior learning is implemented and validated.

5.6.1 Task Environment

Our experimental environment for object-moving is a square arena $\mathcal{R}$ of $100units \times 100units$. If an ant or an object hits the boundary of the environment, the experiment will be reset. At the beginning, ants and the object are randomly placed inside $\mathcal{R}$. The task of the three ants is set to move the object to a desired goal location. The location of the goal is also randomly selected at the beginning. Whenever the object reaches the goal, a new goal location will be selected.

5.6.2 Adaptation to New Goals

While observing the evolution of collective object-moving behavior in ants, we will also study the relationship between group behavior learning for reaching one goal location and that for a different goal. In order to do so, we will conduct an experiment, in which the ants are required to move the object toward to a new goal location as soon as the object reaches its old goal location. We are interested to see whether or not the ants can adapt to the change of a desired goal location by re-selecting some previously-evolved movement strategies.

5.6.3 Simulation Results

In what follows, we will provide some results obtained from our simulation experiments on collectively moving round and square objects with a group of three ants. The specific parameters set in the experiments are

Table 1. A list of parameters set in the experiments.

parameter	symbol	value
number of ants		3
number of contact segments	$\mathcal{N}$	32
environment dimension		100×100
translation step coefficient	α_1	0.5
rotation step coefficient	α_2	$\frac{\pi}{6}$
radius of a round object	R_0	2
dimention of a reactangular-shaped ant		1×1.5
length of a binary chromosome	$bits$	30
population size at each generation	$\mathcal{P}$	8
probability of crossover	p_c	0.8
probability of mutation	p_m	0.05

given in Table 1.

Case 1: Moving a Round Object

Figure 12 presents a series of snapshots showing the movement steps of three ants in moving an object toward a desired goal location. Figures 13 and 14 show the trajectories of the ants and the object. In the figure, the solid line denotes the trajectory of the object, and the dashed line, dotted line and dash-dot line correspond to those of the three ants. Figure 15 shows the corresponding fitness value changes during group behavior learning.

We note that at the beginning of object-moving, ants moved in a rather chaotic manner. The reason is that the chromosomes in a population were just randomly initialized, and no high-fitness configuration was found with genetic operations. It took more than 30 generations for the group of distributed ants to emerge more effective group behavior. This is also reflected in the fitness value changes as shown in Figure 15. The motion trajectories were getting increasingly smoother, as the evolution for collective object-moving continued. After about 40 generations, as shown in the figures, the coordination among the ants became most apparent.

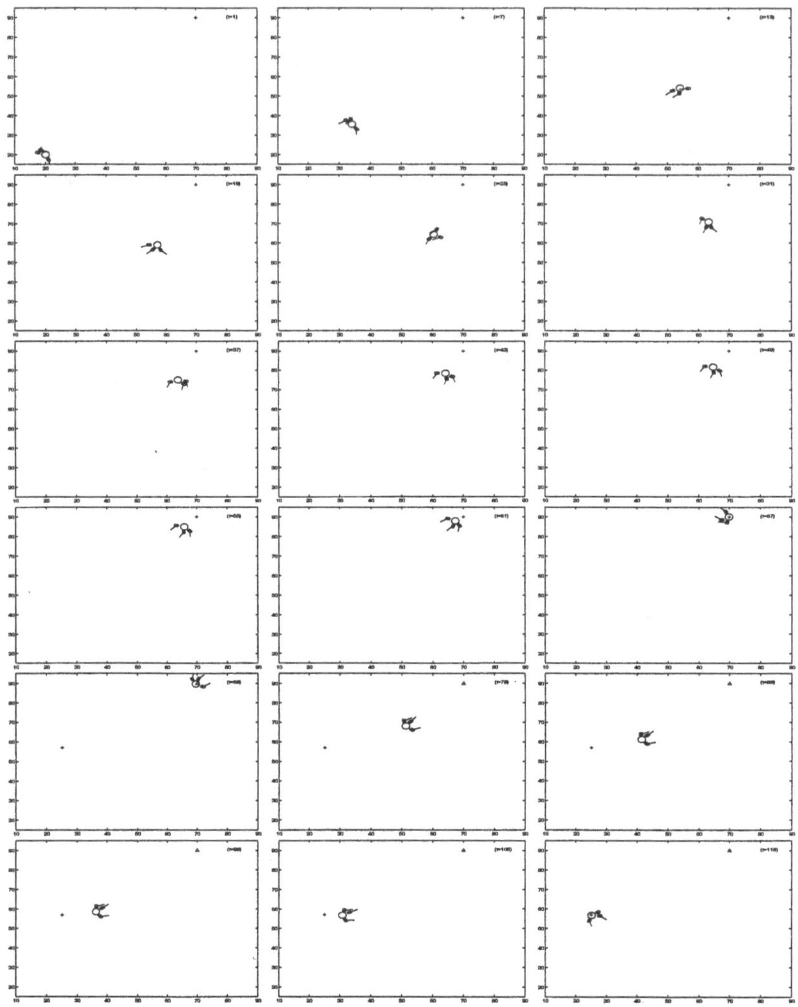

Figure 12. The snapshots of moving a round object by three ants (top view). The collective movement in each snapshot is generated based on *the fittest member of a population* at a certain generation of collective behavior evolution. In the ant environment, a desired goal location is marked with symbol '+'. The number shown at the upper right corner of each snapshot corresponds to the generation number. The tail originated from each ant indicates the direction of its movement. As shown in the figure, good collective movement strategies were found after about 30 generations. At generation 67, the desired goal location was reset. In such a case, previously-found movement strategies were quickly re-selected.

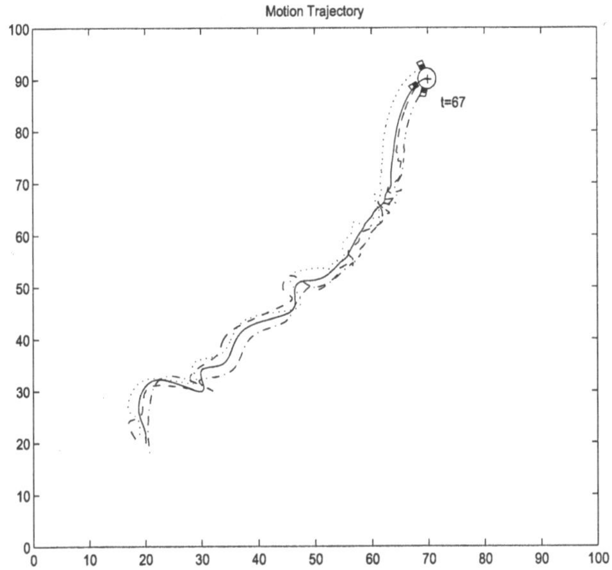

Figure 13. The trajectories of round object-moving by ants from the beginning until the object reached a desired goal location. The solid line corresponds to the trajectory of the object, whereas the others correspond to the movement traces for the three ants.

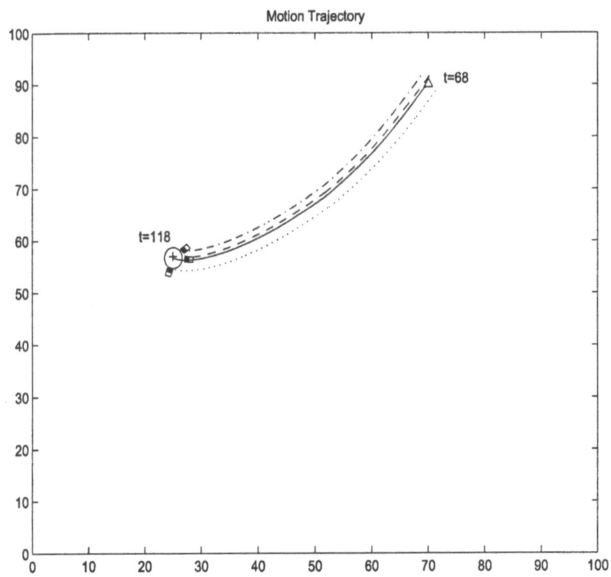

Figure 14. The desired goal location was reset at generation 67.

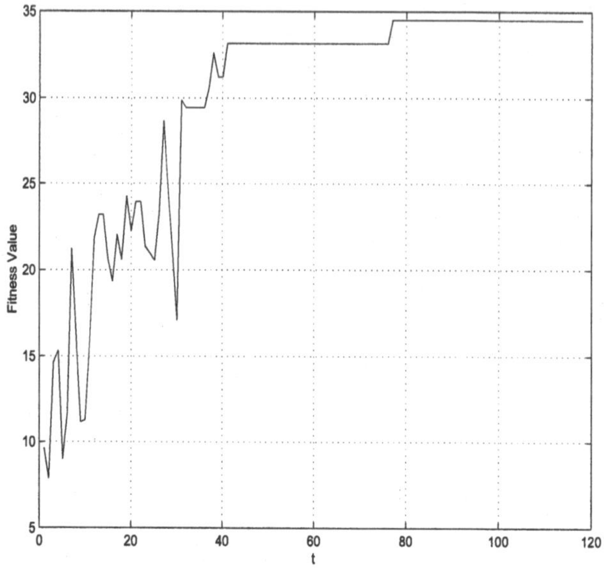

Figure 15. A fitness function plot for *the fittest member of a population* at each generation over the entire collective behavior evolution of 118 generations. Unlike the previous illustrative example, in this experiment, changes in the movements of three ants can be quickly stabilized after about 30 generations, where further recombinations and mutations of the potential movement strategies will not result in significant fluctuations in the fitness values. This is primarily because the present fitness function, e.g., the direction and/or the magnitude of the net pushing force, was not as sensitive as the previous one to the same amount of movement variations.

Case 2: Moving a Square Object

In the case of square object-moving, all parameters, except the object shape and dimensions, are the same as those in the previous case. The side of the square object is 4-unit long.

Figure 16 shows the snapshots of square object-moving. Figures 17 and 18 give the trajectories traced in the process. Figure 19 presents a plot of the corresponding fitness value changes in behavior evolution.

Similar to the previous experiment, the collective movements of three ants can be quickly stabilized after about 20 generations, where further recombinations and mutations of the potential movement strategies will

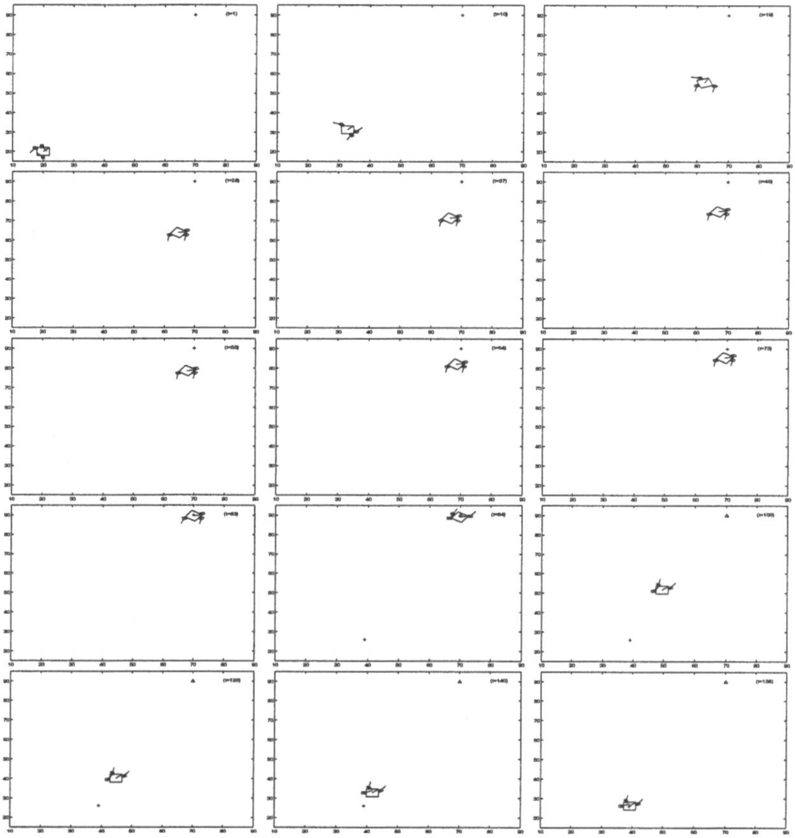

Figure 16. The snapshots of moving a square object by three ants (top view). Similar to the preceding experiment, the collective movement in each snapshot is generated based on *the fittest member of a population* at a certain generation of collective behavior evolution. Except the change of an object, all the rest of notations here remain the same as before. As shown in the figure, good collective movement strategies were found after about 20 generations. At generation 84, the desired goal location was reset. In such a case, previously-found movement strategies were re-adjusted.

not result in significant changes in the fitness values. However, unlike the round object-moving experiment, in the present experiment, after the goal location was reset, the ants cannot simply re-select the previously-found movement strategies due to the orientation of the object created at the time when a goal was first reached at generation 83.

Figure 20 provides a series of object positions and orientations, showing

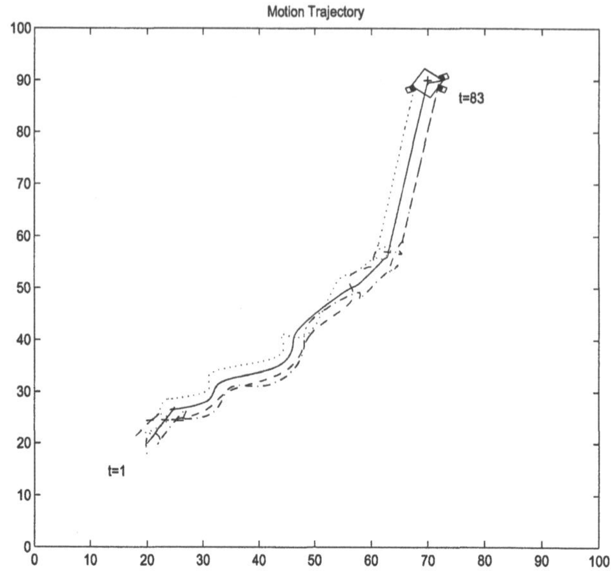

Figure 17. The trajectories of square object-moving by ants from the beginning until the object reached a desired goal location. The solid line corresponds to the trajectory of the object, whereas the others correspond to the movement traces for the three ants.

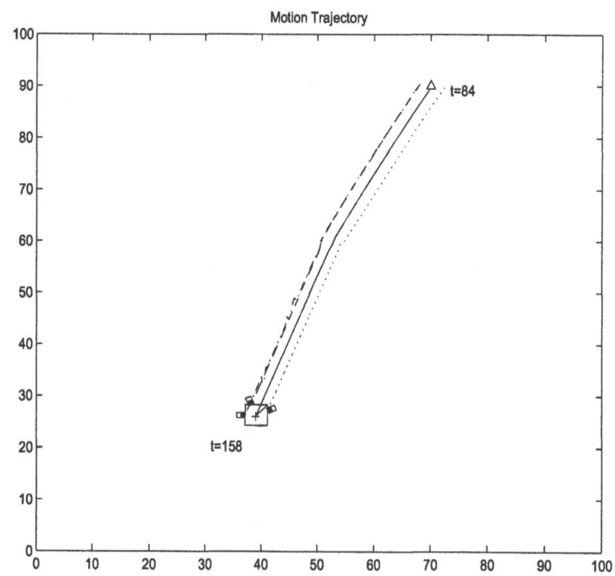

Figure 18. After generation 83, a new goal was set.

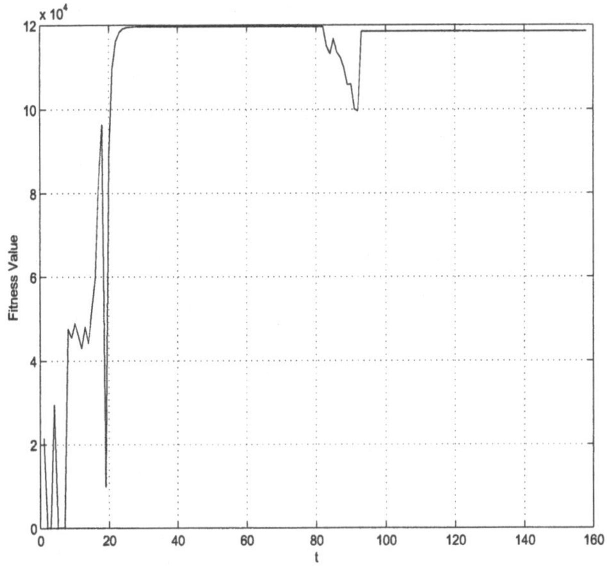

Figure 19. A fitness function plot for *the fittest member of a population* at each generation over the entire evolution of 158 generations. Similar to the previous experiment, the collective movements of three ants can be quickly stabilized after about 20 generations, where further recombinations and mutations of the potential movement strategies will not result in significant changes in the fitness values. However, unlike the round object-moving experiment, in the present experiment, after the goal location was reset, the ants cannot simply re-select the previously-found movement strategies due to the orientation of the object created at the time when the goal was first reached at generation 83.

the translation and rotation of the object as a result of collective pushing. We note that the object was pushed without much rotation (The diagonal inside the object signifies the orientation of the object).

5.6.4 Adaptation to Dynamically Changing Goals

In the preceding experiments, a new goal location was set as soon as ants brought an object to a desired goal location. Here, an interesting question to ask is what will happen if the goal location is changed before it is reached by the object.

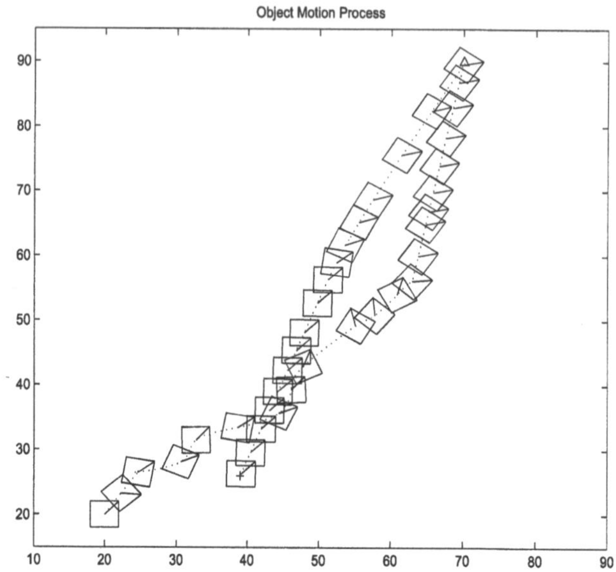

Figure 20. The trajectory of a square object having collectively pushed by three ants from the beginning until the object reached the first desired goal location, and then moved toward a new goal. The diagonal inside the square object signifies the orientation of the object.

Case 1: Moving a Round Object

Figure 21 shows a series of snapshots from collective round object-moving where a new goal location, denoted by '+', was set at generation 81, before an old one, denoted by '◇', was reached. In this case, the ants successfully re-adapted their collective movements. The process of collective pushing before generation 68 has been given in Figure 12.

Figure 22 gives the trajectories of collective round object-moving where the solid line corresponds to the trajectory of the object and the others correspond to those of the three ants. The successful re-adaptation of the ants is also reflected in the smoothness of the observed trajectories. The process of collective pushing before generation 68 can be found in Figure 13.

Figure 23 provides a fitness function plot corresponding to the movement selection and re-adaptation of three ants, as shown in Figure 22, over the

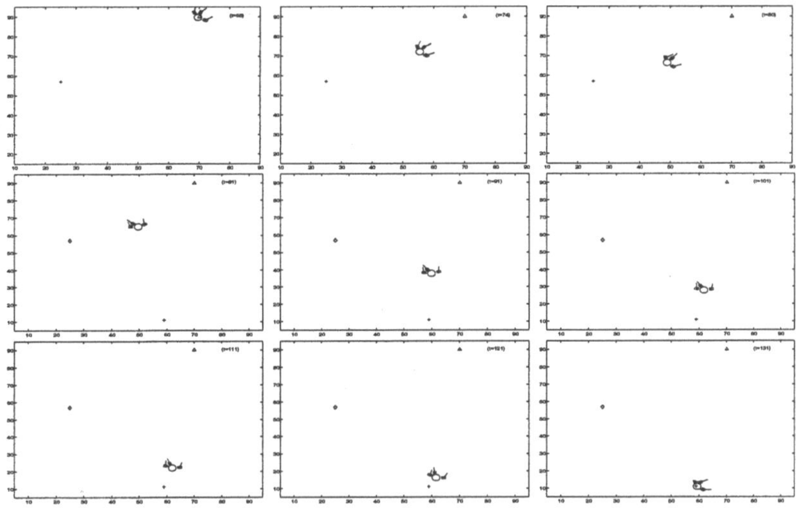

Figure 21. The snapshots of collective round object-moving in which a new goal location, denoted by '+', was set at generation 81, before an old one, denoted by '◇', was reached. In this case, the ants successfully re-adapted their collective movements. The process of collective pushing before generation 68 has been shown in Figure 12.

entire evolution process of 131 generations.

Case 2: Moving a Square Object

Figure 24 presents several snapshots of collective square object-moving. In this case, a new goal location, denoted by '+', was set at generation 91, before an old one, denoted by '◇', was reached. Like the preceding experiment, the ants again successfully re-adapted their collective movements. The process of collective pushing before generation 84 can be found in Figure 16.

Figure 25 shows the trajectory of a square object having collectively pushed by three ants. The successful re-adaptation of the ants is reflected in the smoothness of the observed trajectory. In the trace, the diagonal inside the square object signifies the orientation of the object.

Figure 26 presents a fitness function plot corresponding to the movement selection and re-adaptation of three ants, as shown in Figure 25, over the

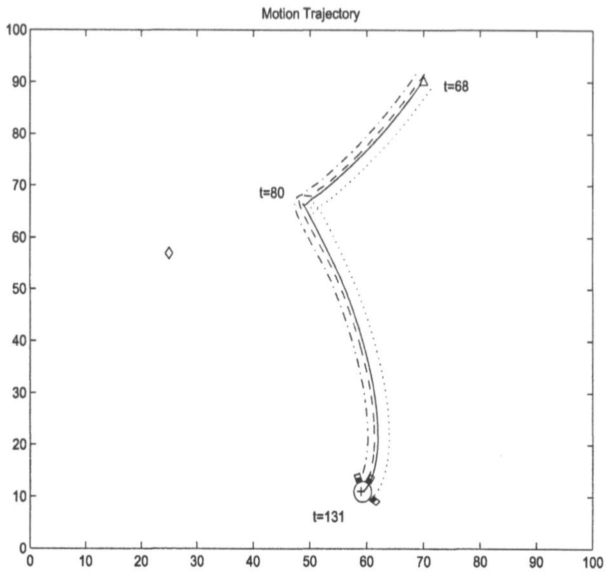

Figure 22. The trajectories of collective round object-moving (the solid line for the object and the others for the three ants). In this case, a new goal location, denoted by '+', was set at generation 81, before an old one, denoted by '◇', was reached. The successful re-adaptation of the ants is reflected in the smoothness of the observed trajectories. The process of collective pushing before generation 68 has been given in Figure 13.

entire evolution process of 152 generations.

5.6.5 Discussions

We can note from the above case studies that ants can quickly adapt to a new goal location based on their early-evolved movement strategies. In the case of round object-moving, the adaptation simply involved the re-selection of a previous collective movement behavior. This is because the orientation of the round object is uniform, and its resulting orientation at the time when it reaches an old goal will not affect the subsequent learning to reach a new goal. In fact, such a case may be viewed as if the ants are undertaking a continuous evolution for collective object-moving toward the same goal location.

On the other hand, in the case of square object-moving, the ants will

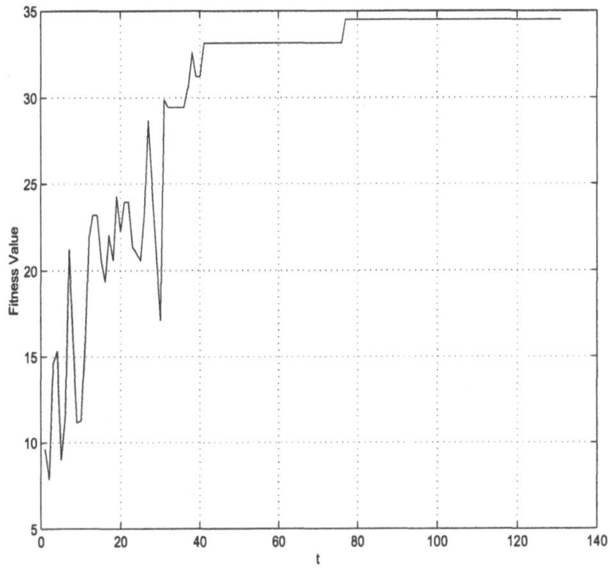

Figure 23. A fitness function plot corresponding to the movement selection and re-adaptation of three ants, as shown in Figure 22, over the entire evolution process of 131 generations.

slightly adjust their movement strategies with respect to the current orientation of the square object, in order to move the object toward a new goal location. This is also reflected in the slight perturbation in the fitness function plot of Figure 26.

In all experiments presented in this section, the object, either round or square, was pushed by the ants through their direct contact forces. Since the size of the object is relatively larger than the one used in the preceding illustrative example of Section 4, the fitness function that measures the direction and magnitude of the net pushing force/torque will be less sensitive to the slight movement variations as compared to the previous illustrative example.

It should be pointed out that the fitness functions used in the experiments are not unique. Other forms of fitness measurements may also be introduced. It is importance to note that the definition of a fitness function can play a role in the resulting performance of collective object-moving behavior learning.

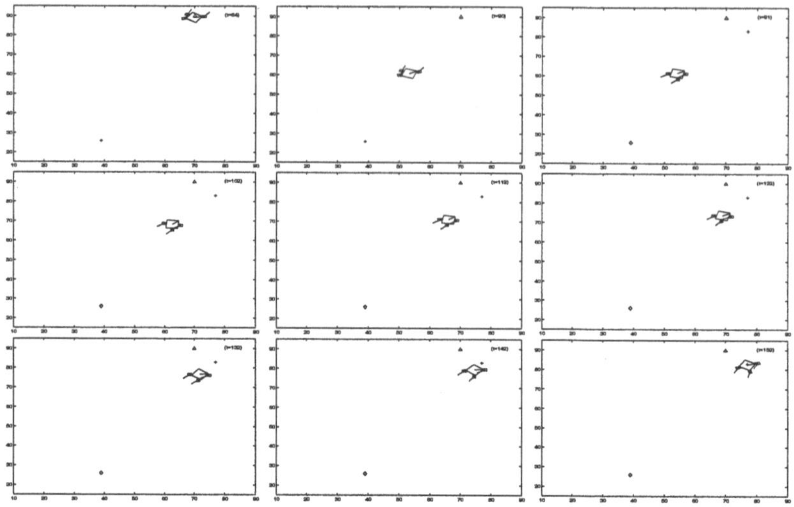

Figure 24. The snapshots of collective square object-moving in which a new goal location, denoted by '+', was set at generation 91, before an old one, denoted by '◇', was reached. Like the preceding experiment, the ants again successfully re-adapted their collective movements. The process of collective pushing before generation 84 has been shown in Figure 16.

All case studies that we have described here are taken from the simulation runs that we have conducted and regarded as being typical. Naturally, this leads us to our next question: From a theoretical point of view, is it possible to estimate the probability of converging to a movement strategy by applying the fittest-preserved genetic algorithm? In the next section, we will explicitly deal with this issue.

6 Conclusions

The object-moving problem may be regarded as a variant of object manipulation problems in material handling, transport, and assembly that require the coordination among a group of distributed ants. In this chapter, we have described a fittest-preserved genetic algorithm-based approach to learning collective object-moving behavior in the ants. We have presented the basic ideas of the proposed approach and its underlying computational architecture, the fundamentals of a genetic algorithm used, and

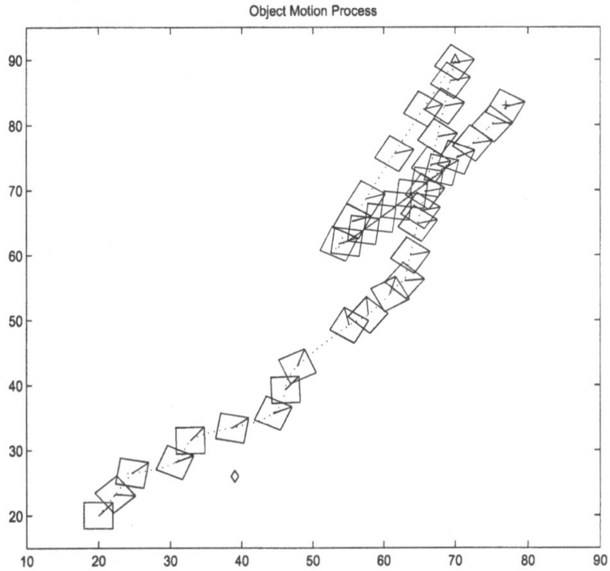

Figure 25. The trajectory of a square object having collectively pushed by three ants. In this case, a new goal location, denoted by '+', was set before an old one, denoted by '◇', was reached. The successful re-adaptation of the ants is reflected in the smoothness of the observed trajectory. In the trace, the diagonal inside the square object signifies the orientation of the object.

the model of the ants.

As in some real-world situations, in our present study we assume that each ant has limited sensing, communication, and on-board computation capabilities. While learning collective movements, it is not necessary for the ants to have exact geometrical and physical models of the object. The ants are required to know only the direction of a desired goal location for estimating their relative orientations.

In order to demonstrate the proposed approach, we have carried out a series of simulations involving ants and objects of various sizes and shapes. The experimental results from some typical case studies have shown that the coupling of local interaction and global goal-directedness created by the ants can effectively select a collective object-moving behavior after a reasonable number of generations.

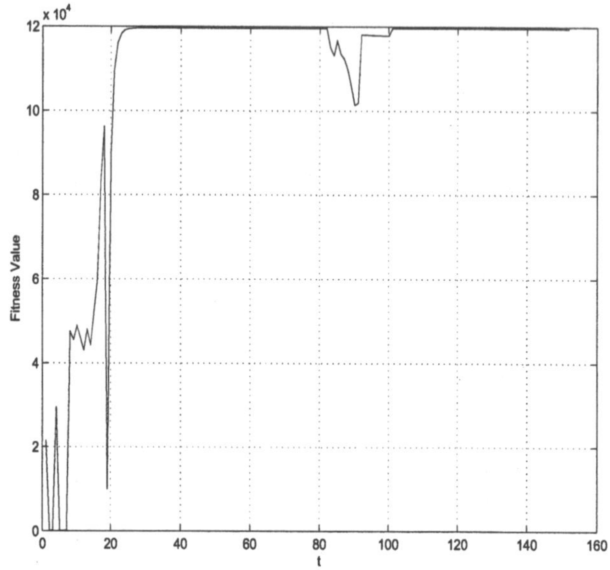

Figure 26. A fitness function plot corresponding to the movement selection and re-adaptation of three ants, as shown in Figure 25, over the entire evolution process of 152 generations.

In order to further examine the adaptation of ants to the changes in goal locations, we have conducted several case studies where the desired goal locations were reset after and before the object had reached an old goal location. We have noted that in both cases the ants can quickly adjust, and sometimes re-selected, previously-found collective movement strategies. Generally speaking, our approach enables the distributed ants to gradually develop an ability of experience-based adaptation to their dynamic task environment.

Acknowledgments

The authors wish to express their gratitude to the editors of this book for their time and great efforts. This research project has been financially supported in part by HK Research Grant Council (RGC) and in part by HKBU Faculty Research Grant (FRG).

References

Back, T., Hammel, U. and Schwefel, H.-P., (1997), "Evolutionary computation: Comments on the history and current state," IEEE Transaction on Evolutionary Computation, vol. 1, no. 1, pp. 3–17.

Balch, T., (1998), Behavioral Diversity in Learning Robot Teams, PhD Thesis, Department of Computer Science, Georgia Institute of Technology.

Beer, R. D. and Gallagher, J. C., (1992), "Evolving dynamic neural networks for adaptive behavior," Adaptive Behavior, vol. 1, no. 1, pp. 91–122.

Bonabeau, E. W., Dorigo, M. and Theraulaz, G., (1999), Swarm Intelligence, Oxford University Press, New York.

Cao, Y. U., Fukunaga, A. S. and Kahng, A. B., (1997), "Cooperative mobile robotics: Antecedents and directions," Autonomous Robots, vol. 4, no. 1, pp. 7–27.

Cao, Y. U., Fukunaga, A. S., Kahng, A. B. and Meng, F., (1995), "Cooperative mobile robotics: Antecedents and directions," in Proceedings of the 1995 IEEE/RSJ International Conference on Intelligent Robots and Systems (IROS '95), pp. 226–234.

Colombetti, M. and Dorigo, M., (1993), "Learning to control an autonomous robot by distributed genetic algorithms," in From Animals to Animats 2: Proceedings of the Second International Conference on Simulation of Adaptive Behavior (SAB-93), pp. 305–312.

Colombetti, M., Dorigo, M. and Borghi, G., (1996), "Behavior analysis and training – A methodology for behavior engineering," IEEE Transactions on Systems, Man, and Cybernetics – Part B: Cybernetics, vol. 26, no. 3, pp. 365–380.

Dorigo M. and Gambardella, L. M., (1997), "Ant colony system: A cooperative learning approach to the traveling salesman problem," IEEE Transaction on Evolutionary Computation, vol. 1, no. 1, pp. 53–66.

Dorigo M. and Gambardella, L. M., (1996), "Ant colony system: A cooperative learning approach to the traveling salesman problem," Tech. Rep. IRIDIA/1996-5, Universite Libre de Bruxelles, Belgium.

Dorigo, M., Maniezzo, V. and Colorni, A., (1996), "The ant system: Optimization by a colony of cooperating agents," IEEE Transaction on System, Man, and Cybernetics-Part B, vol. 26, no. 1, pp. 1–13.

Fujji, T., Asama, H., von Numers, T., Fujita, T., Kaetsu, H. and Endo, I., (1996), "Co-evolution of a multiple autonomous robot system and its working environment via intelligent local information storage," Robotics and Autonomous Systems, vol. 19, no. 1, pp. 1–13.

Fukuda, T. and Iritani, G., (1995), "Construction mechanism of group behavior with cooperation," in Proceedings of the IEEE/RSJ International Conference on Intelligent Robots and Systems.

Goldberg, D. E., (1989), Genetic Algorithms in Search, Optimization, and Machine Learning, Addison-Wesley publishing company, Reading, MA.

Harvey, I., Husbands, P., Cliff, D., Thompson, A. and Jakobi, N., (1996), "Evolutionary robotics: the sussex approach," Robotics and Autonomous Systems.

Holland, J. H., (1975), Adaptation in Natural and Artificial Systems, The University of Michigan Press, Ann Arbor.

Husbands, P., Cliff, D. and Harvey, I., (1996), "The artificial evolution of robot control systems," in Genetic Algorithms in Design Optimization: Proceedings of a Colloquium at The Institution of Mechanical Engineers.

Husbands, P., Harvey, I. and Cliff, D., (1995), "Circle in the round: State space attractors for evolved sighted robots," in The Biology and Technology of Intelligent Autonomous Agents, Luc Steels, Ed., pp. 222–257. Springer-Verlag, Berlin.

Koza, J. R., (1992), Genetic Programming: on the Programming of Computers by Means of Natural Selection, The MIT Press, Cambridge, MA.

Maniezzo, V., Colorni, A. and Dorigo, M., (1994), "The ant system applied to the quadratic assignment problem," Tech. Rep. IRIDIA/1994-28, Universite Libre de Bruxelles, Belgium.

Mataric, M. J., (1997), "Reinforcement learning in the multi-robot domain," Autonomous Robots, vol. 4, no. 1, pp. 73–83.

Mataric, M. J., (1994), "Learning to behave socially," in From Animals to Animats 3: Proceedings of the Third International Conference on Simulation of Adaptive Behavior (SAB-94).

Mataric, M. J. and Cliff, D., (1996), "Challenges in evolving controllers for physical robots," Robotics and Autonomous Systems, vol. 19, no. 1.

Mataric, M. J., Nilsson, M. and Simsarian, K., (1995), "Cooperative multi-robot box-pushing," in Proceedings of the 1995 IEEE/RSJ International Conference on Intelligent Robots and Systems (IROS '95).

Miglino, O., Lund, H. and Nolfi, S., (1997), "Evolving mobile robots in simulated and real environments," Artificial Life.

Nolfi, S. and Parisi, D., (1996), "Evolving non-trivial behaviors on real robots: An autonomous robot that picks up objects," Tech. Rep., Institute of Psychology, National Research Council, Italy.

Reynolds, C., (1994), "Evolution of corridor following in a noisy world," in D. Cliff, P. Husbands, J.-A. Meyer, and S. W. Wilson, Eds., From Animals to Animats 3: Proceedings of the Third International Conference on Simulation of Adaptive Behavior (SAB-94), pp. 402–410, MIT Press/Bradford Books, Cambridge, MA.

Schwefel, H.-P., (1995), Evolution and Optimum Seeking, John Wiley & Sons, Inc., New York.

White, T., Pagurek, B. and Oppacher, F., (1998), "ASGA: Improving the ant system by integration with genetic algorithms," in Proceedings of the Symposium on Genetic Algorithms (SGA'98).

Yamauchi, B. and Beer, R. D., (1994), "Integrating reactive, sequential, and learning behavior using dynamical neural networks," in D. Cliff, P. Husbands, J.-A. Meyer, and S. W. Wilson, Eds., From Animals to Animats 3: Proceedings of the Third International Conference on Simulation of Adaptive Behavior (SAB-94), pp. 382–391, The MIT Press/Bradford Books, Cambridge, MA.

Chapter 8

Applications of Information Agent Systems

M. Klusch, H.-J. Bürckert, P. Funk, A. Gerber, and C. Russ

Intelligent information agents are autonomous computational software entities that are especially meant (1) to provide a pro-active resource discovery, (2) to resolve information impedance of information consumers and providers, and (3) to offer value-added information services and products. These agents are supposed to cope with the difficulties associated with the information overload of the user preferably just in time. In this chapter we describe selected examples of systems of information agents for applications in telematics, e-business, supply chain and repository management. These systems have been developed and implemented by the research group on multi-agent systems at the German Research Center for Artificial Intelligence.

1 Introduction

An *information agent* is an autonomous, computational software entity (an intelligent agent) that has access to one or multiple, heterogeneous and geographically distributed information sources, and which pro-actively acquires, mediates, and maintains relevant information on behalf of users or other agents preferably just-in-time (Klusch, 1999; Klusch, 2001). Thus, any information agent is supposed to satisfy one or multiple of the following requirements.

1. *Information acquisition and management.* An information agent is capable of providing transparent access to one or many different information sources. It may retrieve, extract, analyze, and filter data, monitor sources, and update relevant information on behalf of its users or other agents. The acquisition of information covers activities such as advanced information retrieval in databases and

purchasing of relevant information on appropriate electronic marketplaces.
2. *Information synthesis and presentation.* The agent is able to fuse heterogeneous data and to provide unified, multi-dimensional views on the relevant information to the user.
3. *Intelligent user assistance.* The agent dynamically adapts to changes in user preferences, the information, and network environment. It provides intelligent, interactive assistance for common users supporting their information-based business on the Internet.

In general, we differentiate between *communication, knowledge, collaboration*, and rather low-level *task skills* of an information agent (cf. Figure 1). In this figure, the corresponding key enabling technologies are listed for each of the different types of skills.

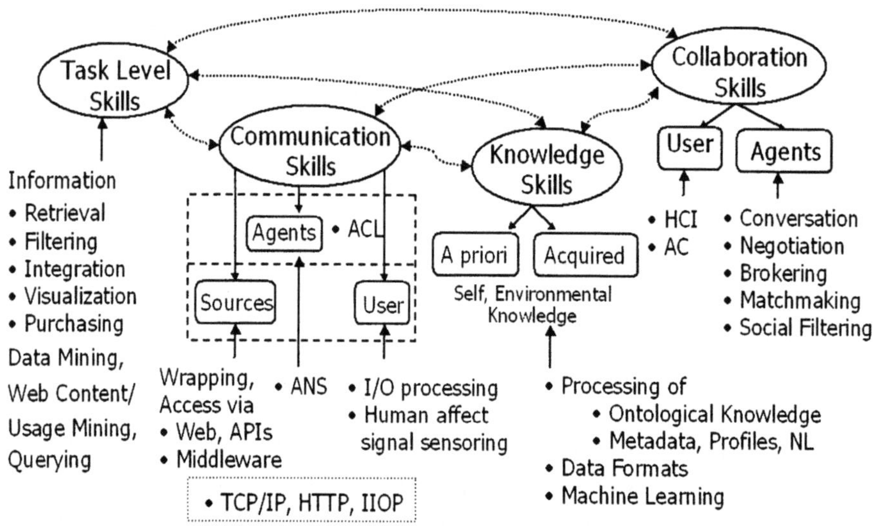

Figure 1. Basic skills of an information agent.

- *Communication.* Information agents communicate either with information systems and databases, human users, or other agents. In the latter case, the use of an agent communication language such as FIPA ACL has to be considered on top of, for example, middleware platforms or specific APIs such as CORBA/IIOP, JDBC (Java Database Connectivity), and Microsoft's ODBC (Open Database

Connectivity) or OKBC (Open Knowledge Base Connectivity), respectively.

- *Knowledge*. An information agent copes with pre-given or dynamically acquired data, information, and knowledge on its environment and itself. This may include representation and processing of ontological knowledge and metadata, profiles and natural language input, translation of data formats as well as the application of machine learning techniques.
- *Collaboration*. The higher level interaction and collaboration with other agents may rely on techniques and protocols such as for service brokering, matchmaking, negotiation, and social filtering. Interaction with human users mainly corresponds to the application of techniques stemming from human-computer interaction (HCI) and affective computing (AC).
- *Task*. The core tasks of any information agent include retrieval, filtering, integrating, and visualizing relevant information. This implies in particular effective utilization of methods for Web content and/or usage mining as well as optimal query planning and processing.

In fact, the extent of utilizing one or multiple of these categories of basic skills varies depending on the application domain and tasks the information agent has to accomplish. For a comprehensive survey on information agent technology for the Internet we refer the reader to (Klusch, 2001).

The remainder of this chapter is structured as follows. In the Sections 2, 3, and 4 we present different applications of holonic-structured information agent systems in the domain of telematics and e-business. Section 2 briefly describe systems for dispatch support (TeleTruck), and remote control and diagnostics of technical systems (TeleService). In Section 3 we survey a system of collaborating information for mobile integrated services for e-trading in selected application domains of forestry and agriculture. Section 4 then proposes a general agent-based coordination infrastructure for retail supply webs in the e-business domain. Finally, Section 5 presents an approach of information agents to support software repositories. Each of these applications have been developed and implemented by the research group on multi-agent systems at the German Research Center for Artificial Intelligence.

2 Holonic Agents for Telematics

2.1 TELETRUCK – a Dispatch Support System

The TELETRUCK application prototype aims at supporting dispatch officers of forwarding companies. It allows for interaction with on-board systems installed in the trucks of the company using satellite or mobile phone communication. Different copies of TELETRUCK can be connected to distributed system for collaborative transport scheduling – used by dispatch officers of collaborating companies or different sites or departments of one company.

We have modeled the basic physical objects (drivers, trucks, trailers, containers, dispatch officer) of the transportation domain explicitly by agents. The agents can join together forming *holonic agents* or *holons* that act in a corporate way: driver, truck, trailer, and container agents compose to a transport holon representing a physical transportation unit (a road train or an articulated vehicle together with its driver), which is able to execute a transport order. These holons are coordinated by a dispatch agent. Together the transport holons and their dispatch agent form again a holon, which acts as planning and scheduling assistant for a human dispatch officer.

Technically the dispatch agent distributes incoming transport orders to transport holons using contract net protocols. The transport holons compute a bid for an announced order on the basis of their current transport plan (for earlier distributed orders). Dispatch agents select the best bid and award the order to that holon. Not every transport holon has to be composed at this stage. Some may still be incomplete, e.g., missing a trailer with further loading space, which could be integrated. Others still have to be composed of idle component agents from the scratch. Composition and extension of transport holons is organized by a cascading sequence of transport protocols (cf. Bürckert *et al.* 2000 for a detailed description.)

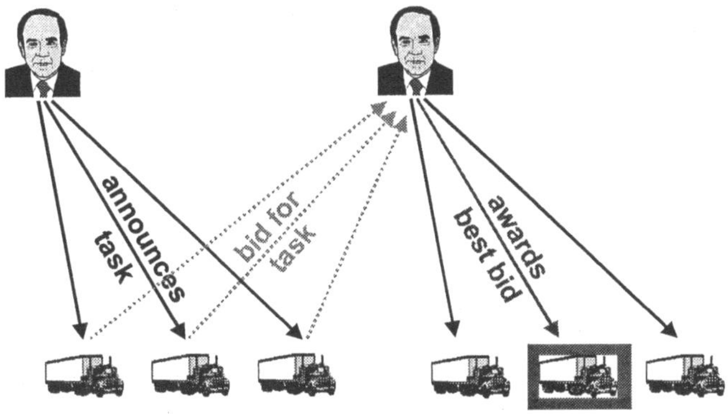

Figure 2. Contract Net Protocol for Order Distribution.

Since such transport plans are computed for a sequential stream of incoming orders only optimizing the local plans of the transport holons, they are suboptimal by nature. Hence the exchange of orders between different transport holons is organized by a simulated trading protocol. Again coordinated by the dispatch agent, transport holons can sell some of their orders and buy others, in order to improve their plans.

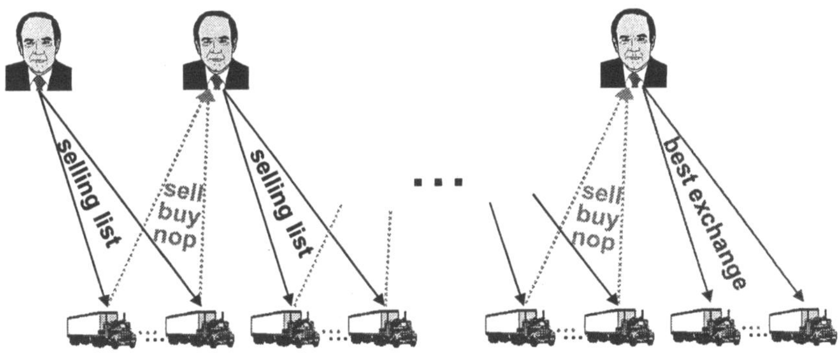

Figure 3. Simulated Trading Protocol for transport plan optimization.

Because of the randomized selection of the orders to be sold or bought, this method results in a simulated annealing like procedure that optimizes the global transport plan of one dispatch officer incrementally. For several connected TELETRUCK systems we use a similar organization of the whole network. One system acts as coordinator – it makes sense to have here a system without transport holons. It announces – again within a contract net protocol – the transport orders to the other systems' dispatch agents, which then in turn announce them to their transport holons as described above. A simulated trading protocol can also be used for optimizing the transport plans globally for the whole network of collaborating systems. However, if we support collaborating companies, we use competitive versions of the contract net and the simulated trading protocol (cf. Vierke 2000).

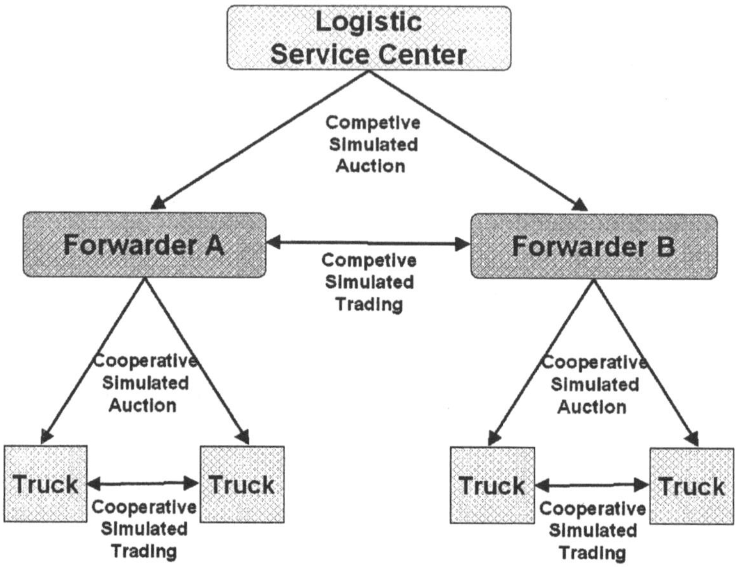

Figure 4. Contract Net and Simulated Trading for collaborating companies.

2.2 TeleService – Mobile Agents for Remote Applications

Remote diagnostics, maintenance, control and other tele-services for stationary or mobile technical systems may become a typical field for mobile agent technology. The basic idea is to support service providers of technical systems, which are distributed – in some cases all over the world, in their daily job. Usually a service center sends out service engineers to its customers – for routine checks or on demand. The engineers check the system or solve the problem and move to their next job. Mobile agents could overtake such service tasks – at least for simple or routine cases – provided the technical systems are connected with the server of the service provider.

In our TeleService approach we have software agents representing each technical system. Control devices make the current state (such as the current location in the case of mobile systems) fully available to these agents and also provide the agents with interfaces for communication and interaction.

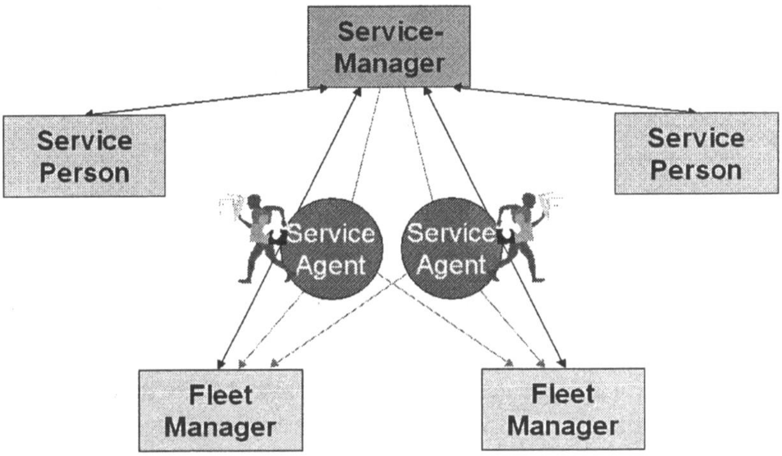

Figure 5. Service agents are sent out by service management agents.

Mobile service agents are used for diagnostics, maintenance and control of the systems. Starting from the server – at a service center or at the customer – they visit local agents and execute their tasks. Where the underlying computing platform is physically mobile, the hosting of agents on the onboard computer is supported.

Currently we are instantiating this approach for forklift trucks which are equipped with control processors and onboard computers. At the service center, an agent coordinates service management for several customers and their forklift fleets. At each customer's server, other agents are responsible for the fleet management. Images of the forklifts are realized as agents which know all about the current status of the forklifts.

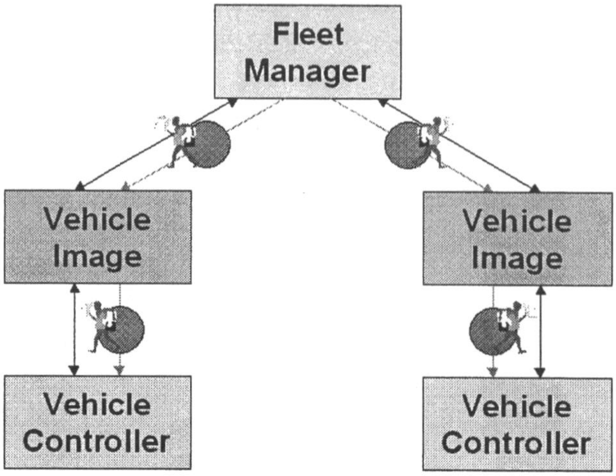

Figure 6. The fleet management agent moves the service agents to vehicle agents which in turn upload the service agents to the board computer.

Service agents for different services can be sent by the service management agent via the fleet management agents to the image agents, which can upload the service agents onto the onboard computers of the forklifts (cf. Figures 5 and 6).

The agent based approach allows for high flexibility and frees the agent from dependence on a permanent data connection between the servers and onboard systems, which would be needed in a conventional architecture for remote applications: as service agents can execute their tasks autonomously a permanent connection is not necessary, even for control tasks. Only uploading of the service agent onto the onboard computer requires a brief data connection. In the reverse direction, only the results of the activity of the service agent are communicated.

Since the onboard computers and control processors are usually of very low computational power, the service agents cannot be realized as big intelligent expert systems but as small agents with very specific expertise. In order to solve, for instance, a complicated diagnostics problem several of these specific agents may be sent one after the other to the onboard system for analyzing different aspects of the failure. Their results are collected by the service management agent which then decides for further activities to repair the failure. Such activities could range from tasks to be done again by service agents over tasks to be solved by technical staff at the customer (e.g. a battery exchange) to even complex tasks that require a service engineer which has to be sent by the service provider.

The intermediate layers (i.e., the fleet manager and the images of the target systems) of the approach allow for secure transfer of the mobile agents provided suitable secure interaction protocols are used. Only the image agents of the target systems are allowed to upload service agents onto the onboard computers.

3 CASA: Agents for Mobile Integrated Commerce in Forestry and Agriculture[1]

3.1 Motivation

The project CASA (Cooperative Agents and Integrated Services for Logistic and Electronic Trading in Forestry and Agriculture) focuses on the development of an agent-based information and trading network (CASA ITN) for mobile, integrated services in selected application scenarios within the domains of forestry and agriculture in the Saarland county. The CASA ITN can be installed at users connected to the Internet; registered users may then access the complete CASA services via a secure Extranet and mobile telecommunication network.

Agent-mediated services of the CASA ITN aim to support the main operative business processes users are performing in each of the following application scenarios: (1) customer-oriented, dynamic timber production, (2) mobile trading of timber via different types of auctions, fixed or negotiable price, and (3) electronic trading of cereals.

The approach taken for providing information and trading services in the CASA ITN focuses on the effective integration of production, logistics and trading processes of respective parts of the particular supply chain for each of the application scenarios.

[1] This research and development is sponsored by the Saarland Ministry of Economics under grant 032000

This approach is motivated by the paradigm of *integrated commerce* (i-commerce) which can be seen as an operational extension of traditional e-commerce. The basic ideas of i-commerce are (a) to get customers more involved in the activities related to his/her orders or tasks to be jointly accomplished by one or multiple contractors, auxiliary/finishing industry, and shipping companies, and (b) to get related processes in the considered supply chain more integrated in practice.

The agent-based CASA service suite for i-commerce can be accessed by users at any time, anywhere; selected information and trading services are available via mobile WAP-enabled computing and telecommunication devices such as smart phones, PDAs and communicators. Efficient coordination of services within the CASA ITN is done via appropriate types of collaborating software agents.

3.2 CASA Agents and Services

3.2.1 Holonic Agent System of the CASA ITN

We differentiate between following groups of participants in the CASA ITN: *producers* offer their goods; *buyers* want to purchase several goods; *retailers* act in agency of a company or of their own profit; *logistics companies* are responsible for transportation tasks, storage and resource management. Each member of these groups is represented in the ITN by an appropriate kind of software agent (cf. Figure 7).

For reasons of effectively accomplishing complex, mostly hierarchically decomposed tasks and resource allocations in the selected application scenarios (cf. section 3.3) we use the concept of *holonic agents* (Gerber, Siekmann, and Vierke 1999) and *holonic multi-agent systems (H-MAS)* (Bürckert, Fischer, and Vierke 1998). A holonic agent (or holon) coordinates and controls the activities and information flow of its subagents. In a H-MAS, autonomous agents may join others to form, reconfigure, or leave a holon.

A human user in the ITN is represented by a special holonic agent called *personal assistant*. It pro-actively acts on behalf of its user even if (s)he is off-line; the personal assistant is the coordinating head of a set of other specialized agents for individual negotiation, participation in auctions, finding other relevant partners for trades and information, and elaboration of optimal trading strategies over time.

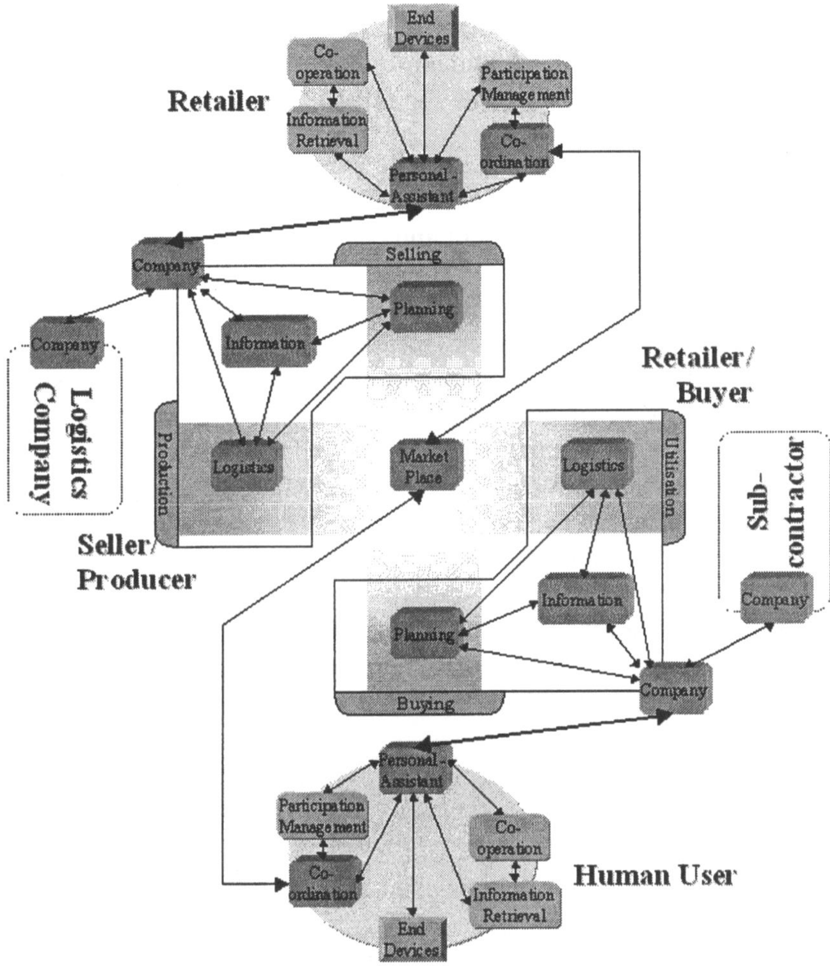

Figure 7. Overview of the holonic CASA agent system.

Each corporation is represented by a special holonic agent system according to its task-oriented subdivision into departments for information management, logistics, and production planning.

In this context we presume that:

- *information management* services provide information either on certain products and related production processes, or on current market situation and potential competitors.
- *logistics* services support the coordination of machines for production and transport, human resources, and storage capacities.
- *production planning* services support short-, middle-, and long-term product planning cycles.

A corporation holon is constituted by other holonic agents each of them representing a special department and corresponding complex tasks and services. Since in the CASA ITN the roles of buyer/retailer or and seller/producer may be interchangeably used both are modeled by similar holonic agent structures. In addition, logistics companies are usually contracted by other corporations for the purpose of time- and cost-saving delivery of goods on demand. Finally, to enable different kinds of trading between the participants of the CASA ITN such as multiple online auctions and sales via fixed or negotiable prices in bilateral negotiations at the same time we developed agent-based services for a distributed virtual market place. This enables each participant to, for example, initiate and perform one or multiple auctions or negotiations for sales of its own goods and products.

3.2.2 Agent-Based Services of the CASA ITN

In general, the CASA agent society provides the following classes of services to its users and is coordinating the effective execution of each of the services.

- *Auction mechanisms* including Dutch, English, Vickrey, and First-Price-Sealed-Bid auctions.
- *Integrated services* for
  - *dynamic pricing*. Agents collect additional information on transportation costs and other constraints to meet as a decision-support service for its users, for example, during the bidding processes of some auctions.
  - *logistics*. These services provide dynamic, approximately optimal (re-)scheduling and (re-)planning of transportation.
  - *information management.* Agents gather relevant information on behalf of its users in different trading and production settings.

- *Mobile services* to let the users access most services of the CASA ITN also on WAP-enabled mobile devices.

3.3 Application Scenarios

In brief, the application scenarios of the CASA ITN are as follows.

- *Customer-oriented dynamic timber production* (KDPS): Foresters and timber harvester appropriately cooperate with pro-active support of service providing agents of the CASA ITN to accomplish the goal of an individual customer's order to deliver a certain quantity of timber with specified quality at a given time of delivery. The approximately optimal, dynamic (re-) planning and coordination of integrated services for harvesting, processing, and transportation is performed by the CASA multi-agent system and partly accessible via mobile WAP-enabled devices.
- *Mobile timber sales* (MHS): Registered users of the CASA ITN may set up and participate in one or multiple timber auctions via mobile WAP-enabled mobile devices.
- *E-trading of cereals*: Similar to the MHS scenario registered users of the CASA ITN may trade grains via (a) auctions (with the farmer being the auctioneer) or (b) multi-lateral negotiations with interested parties following a version of the extended contract net protocol. Many of the required services in this scenario can be appropriately re-used from those developed in the MHS and KDPS scenario

The first two of these application scenarios have been implemented in the CASA ITN.

3.3.1 Mobile Timber Sales: Services, Interactions, and Agents

In this special scenario each ranger may sell timber via different kinds of auctions, fixed or negotiable sales offers to other registered users of the CASA ITN. Main benefits of the agent-based service support are the concurrent monitoring and computation of optimal transport costs per individual bid or buying offer, and the full mobile service support via WAP-enabled mobile devices.

In general, the mobile timber sales services of the CASA ITN enable registered users (a) to initiate and/or participate in one or multiple timber auctions, and (b) to sell or buy timber at fixed or negotiable prices. In the first case, any user may set up an auction thereby acting as an auctioneer of (parts of) his/her own goods for sale. The CASA ITN currently provides the following types of auctions: Dutch, English, Vickrey, and First-Price-Sealed-Bid. There is no central market place. An auctioneer may either be identical with the seller of some goods, or different. In the latter case, an auction could take place at a different but trusted site on behalf of the seller which in turn may increase the trust of potential buyers of the auction goods. The auction server has been built upon a general holonic coordination server (cf. Section 4.5.2; Gerber and Russ, 2000). Any user may invoke *integrated services* for decision-support during the participation in one or multiple different auctions. For example, a personal CASA agent may concurrently determine the optimal transportation costs and delivery dates of some auction good for each individual bid of its user. As a result, the agent may notify its user in real-time if estimated optimal transport costs exceed the allowed limit due to given buying preferences or if some deadlines are at risk to be exceeded.

The integrated trading services are provided by CASA agents to support interaction between the participants in the scenario, that is seller and/or auctioneer, buyers, and shipping companies. Figure 8 summarizes such interaction in case of an auction. First a seller initiates an auction on a trusted auction site and informs potential buyers about offered goods, the auction type, and related bidding policy. During an auction bidders may evaluate their bids, monitor the current auction process, and place bids according to their individual bidding strategy. Latter may evolve depending on the result of the integrated services and respective benefits for the individual bidder.

In addition, each of the information and trading services is available for registered users of the CASA ITN on mobile WAP enabled devices such as smart phones or PDAs connected to the internet via modem. A synchronization between the mobile and PC-based home computing environment in the CASA ITN is coordinated by appropriate CASA agents. Participation in any trading process can be delegated by the user to its personal agent which then is in charge of negotiating or bidding at an auction, and notifying its user when needed, e.g., via SMS or email.

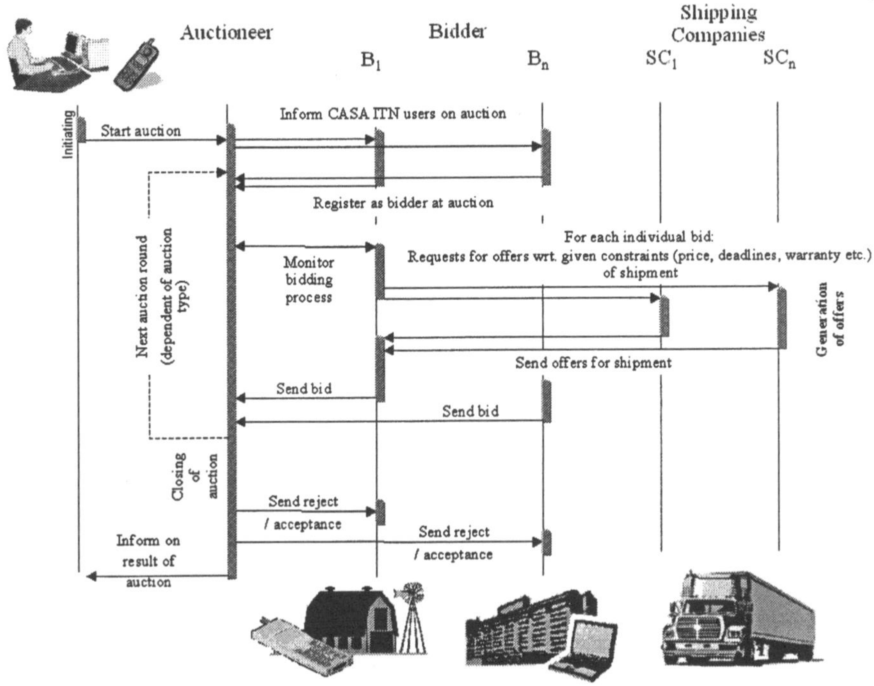

Figure 8. Interaction between participants in the MHS scenario.

3.3.2 Relevant Holonic Agents in the MHS Scenario

Figure 9 shows the different kinds of holonic agents which play an active role in the MHS scenario. These are holons for users as buyers or sellers/auctioneers, and shipping companies. We distinguish between buyers with or without logistics capabilities. In the latter case, carriers have to be appropriately contracted by the buyer. A company holon is structured into three different department holons as mentioned above.

3.3.3 Implementation

The services of the CASA ITN have been implemented in Java (JDK 1.3); the prototype of the system has been extensively tested on Windows NT/ME/2000 machines. CASA holonic agents are modeled in accordance with the InteRRaP agent architecture and implemented using the standard open source FIPA-compliant agent system development environment FIPA-OS 2.0. Implementation of the mobile application services of the CASA ITN is compliant with the WAP 1.1 standard. These services are accessible by users of the current version of the ITN via the secure T-D1 WAP-gateway of the Deutsche Telekom AG. Data security is provided by the virtual private network suite of Checkpoint AG.

Figure 9. Relevant holons in the mobile timber sales scenario.

4 MAS-R/3: A Multi-Agent Coordination Infrastructure for Retail Supply Webs[2]

4.1 Motivation

According to Booz-Allen & Hamilton, in retail industry as well as in many other industries supply structures are evolving which may be

[2] This work has been sponsored by the SAP AG subsidiary SAP Retail Solutions.

characterized as "*complex sets of relationships that appear more web-like than chain-like*". Such structures are called *supply webs* (Laseter 1998). Partnerships between autonomous business entities in these supply webs can be flexibly contracted or withdrawn and are predominantly short-dated. This may cause complex coordination problems since the resulting many-to-many interactions and instantiated supply paths are not stable but may dynamically change. The coordination of the manifold interactions and internal planning tasks overstrains present-day inventory control systems.

We tackle these coordination problems by applying the concept of intelligent agents to the design of retail supply webs. Each operative unit of a retail supply web is modeled as an agent which represents either a supplier, producer, broker, wholesaler, retailer, warehouse, distribution center, branch, or logistics service provider. The coordination infrastructure for these *supply web agents* relies on *coordination agents, services, and mechanisms* such as auction server agents. Latter may provide other services such as matchmaking services, and auction types, coalition forming mechanisms for (re-) allocation of resources, respectively

In subsequent sections we briefly describe the considered application domain, our proposed design for a warehouse agent as an example of a supply web agent, and a short discussion of the implemented coordination server.

4.2 The Supply Web Application Domain

The supply web application domain can be structured into three different layers each of them representing a different view on the domain (cf. Figure 10). The first layer represents the *supply web structure*. As in any supply chain available resources are (a) transformed to goods or products along some supply paths, (b) allocated to brokers, wholesalers and/or retail concerns, and finally (c) distributed by them (potentially via additional intermediators) to the consumers. However, in modern competitive settings of electronic markets supply paths are becoming more and more established quite short-dated, that is not until a real supply flow between appropriate supply web entities is impending. This implies the need to flexibly exchange parts of supply paths. The situation gets even more complex in cases where the outcome of the competition of supply web entities for the assignment of goods in the supply chain is not known in advance (see circle in Figure 10). Depending on who wins this competition all of the subsequent entities in the "winning" supply path are enforced to adapt their plans correspondingly. The second layer models what we can consider as the *agentification view*, i.e., how a set of supply web agents is mapped to corresponding "real" entities within the supply topology.

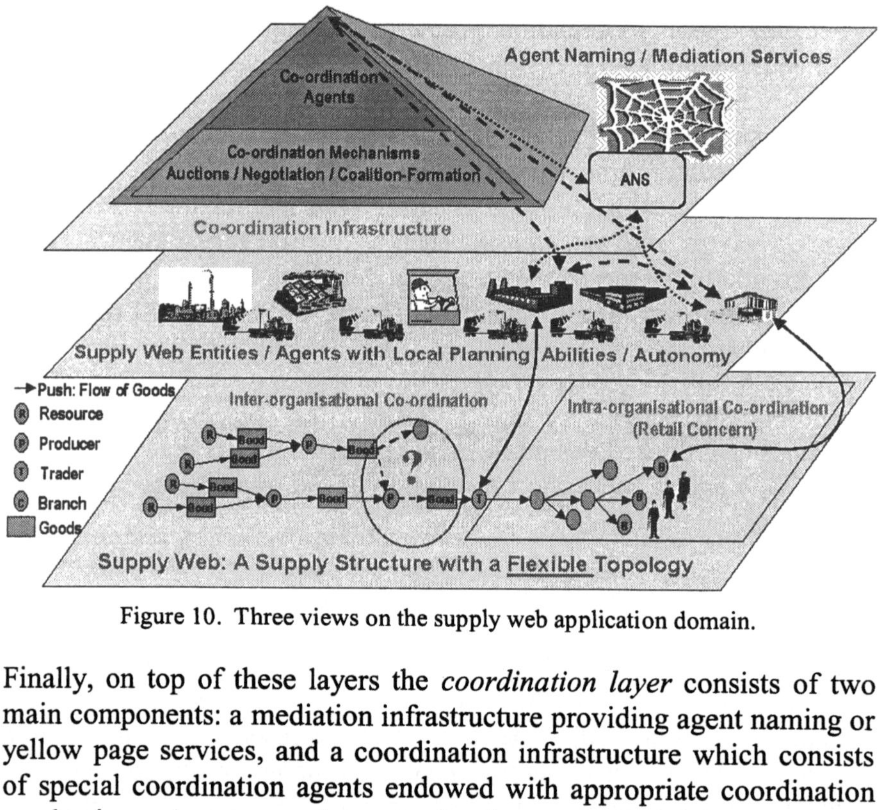

Figure 10. Three views on the supply web application domain.

Finally, on top of these layers the *coordination layer* consists of two main components: a mediation infrastructure providing agent naming or yellow page services, and a coordination infrastructure which consists of special coordination agents endowed with appropriate coordination mechanisms. Agents may interact after they registered themselves at an agent name server (ANS).

4.3 Agentification of Supply Web Entities

How can supply web entities be agentified, how can the respective supply web agents be modeled in our scenario? We have implemented the operative supply chain units in Java by using the proprietary agent development tool *gACE* (generic Agent Control Engine) (gACE). *gACE* agents are equipped with a high level control architecture, called *Procedural Reasoning System (PRS)* (Ingrand et al. 1996), which allows the agents to behave goal-directed while staying responsive to their environment. Agents may execute predefined basic actions and given goals represented by plans on stack (out of the agent's plan library) as well as to replan their actions. gACE's basic modules are the *agent control engine (ACE), the agent control unit (ACU), a knowledge base* and a *communication unit* (cf. Figure 11).

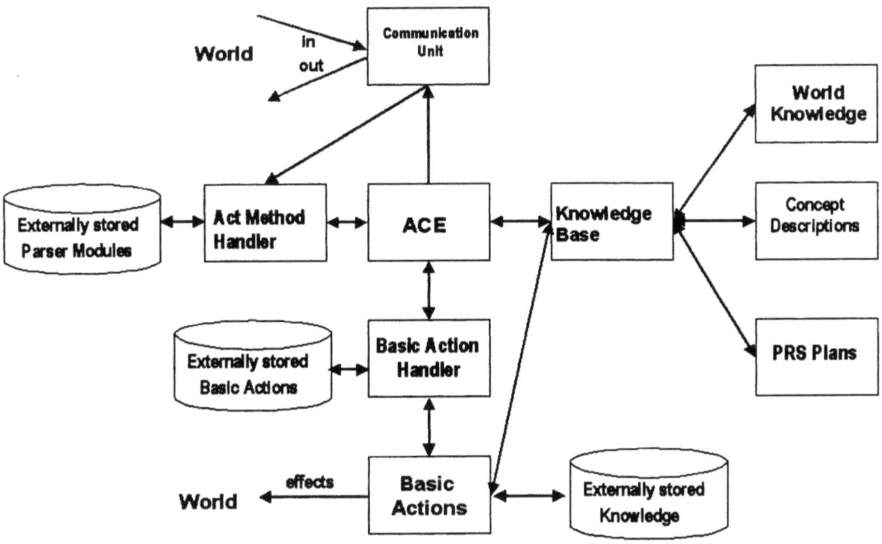

Figure 11. Architecture of the gACE framework.

The *Agent Control Engine (ACE)* connects all parts of an agent and controls the ACUs spawned at runtime for controlling the processing of activated plans. All incoming and outgoing messages are handled by the *communication unit* using the UDP protocol for data transmission and the KQML language as message format. Incoming messages are parsed by the *Act Method Handler,* the knowledge base is updated, and plans as sequences of selected basic actions are instantiated and activated (*Basic Actions Handler*).

4.4 Warehouse Agents

The holonic warehouse agent is an example of a special supply web agent (cf. Figure 12) and has been implemented with *gACE* .

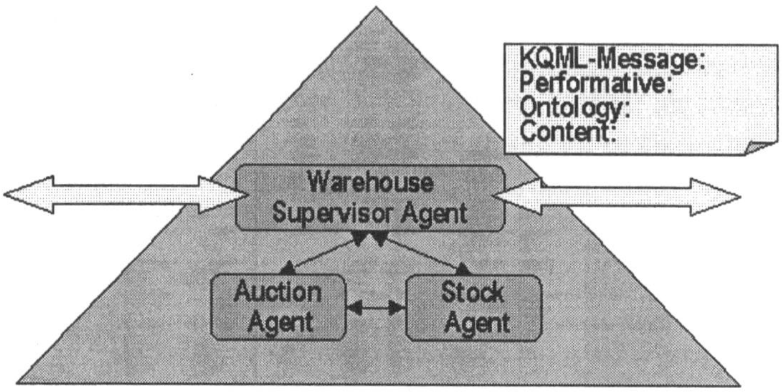

Figure 12. The Warehouse Agent Holon.

This holon consists of different types of interacting agents: The *supervisor agent* represents the interface to the supply web environment and coordinates the operation of its subagents; the *auction agent* has to handle all activities related to auctions; the *stock agent* manages the flow of goods as well as policy based reservations.

4.5 Supply Web Coordination

4.5.1 Coordination Policies

The warehouse agent behaves according to given rules (policies) of which the main ones are the following.

- *Supervisor Policy.* The supervisor agent decides on which good has to be traded at what time and for what price. For example, the *keep-the-level policy* describes the re-ordering of goods in case of the entity falls short of storage capacities.
- *Bidding Policy.* The auction agent determines the value of the bids. For example, the *always-one-higher policy* prescribes a

continuous increase in bid value by one until a certain limit is reached.

- *Decommitment Policy.* Any commitment or reservation of resources may be cancelled with penalty. Various policies could manage the decision if and when to decommit.
- *Reservation Policy.* An agent has to decide on if some proposed reservation of goods for another agent is acceptable, or not.

Other behavior-oriented policies include policies for supply paths, and calculation of delivery dates and delays. Latter kind of policies can be categorized into *just-in-time policies, safety lead-time policies, bid refusal policies,* and *promise date negotiation policies.* A comprehensive overview of these types of policies can be found in (Kjenstad, 1998).

4.5.2 The Supply Web Coordination Server

The coordination server has been structured into a 3-layered holonic agent architecture as follows (see Figure 13).

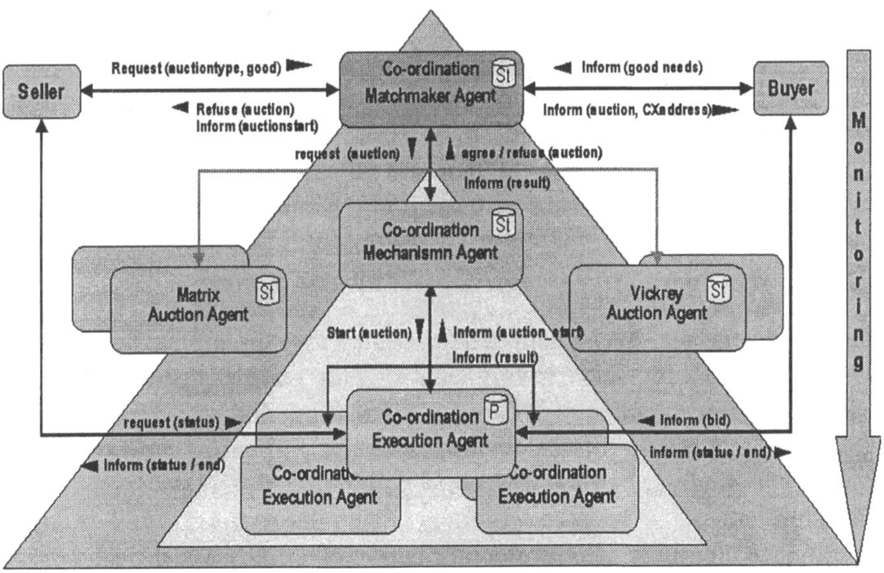

Figure 13. The Supply Web Coordination Server.

The Coordination Matchmaker Agent (CMM Agent) serves as an interface to supply web agents which are requesting for coordination of their sales activities in terms of an auction. For each request the CMM agent selects the most appropriate coordination mechanism (CMech) agents for further processing of the request. In case there are enough resources available for initiating an auction of the desired type by the subordinated coordination execution (CX) agents the respective supply web agent(s) will be notified on the details of this special auction matching its request. If the request matches with some currently running auction(s) the CMM agent forwards the contact details of the corresponding CX agent(s) it receives by the respective CMech agent(s) to the supply web agent.

Latter can then directly contact the selected CX agent(s) for bidding or status monitoring depending on its role as buyer or seller, respectively. The CMM agent is in charge of efficiently coordinating and monitoring operative processes of running auctions to ensure appropriate load balancing at the server. For reasons of statistics the result of each auction is also propagated internally in a bottom-up fashion to each of the responsible agents, i.e., CX, CMech, and CMM agent.

4.5.3 Market-based Supply Web Coordination Mechanisms

We have endowed our supply web coordination server with several market-based allocation mechanisms for the coordination of supply web activities such as the simulated trading algorithm (Bachem et al., 1992) and matrix auction (Gomber et al., 1998).

The *simulated trading (ST) algorithm* is a randomized algorithm that realizes a market mechanism where contractors attempt to optimize a task allocation by successively selling and buying tasks in several rounds. *Matrix auctions (MA)* are truth-revealing and - in contrast to ST and VA - applicable for the *simultaneous* assignment of multiple items or tasks to bidders. In a *matrixkauction (MAk)* ,*k* items are auctioned-off simultaneously to some bidders. From their transmitted bids an auctioneer identifies the optimal allocation of all *k* items. Prices are set according to the *Vickrey auction (VA)* (Vickrey, 1961), i.e. bidders receiving items pay the second-highest bid made for these items.

Empirical results on the suitability of these coordination mechanisms for the allocation of transportation tasks in a network of shipping companies are reported in (Gerber et al., 1999); the mechanisms ST, MA-2, and MA-3 showed promising performance for supply web coordination tasks.

4.6 Future Work on MAS-R/3

We have implemented the coordination infrastructure and supply web agents in Java. Future efforts will mainly be devoted to the integration of additional coordination mechanisms into the infrastructure. This aims to decentralize the configuration and coordination of distributed business processes as well as the (re-)allocation of resources and tasks within complex supply webs.

5 Agent-Based Support of Software Repositories[3]

5.1 Motivation

The availability of knowledge and information about internal business processes and software systems is strategically crucial and relevant for companies. One means to make such knowledge available in a smooth and easily usable way are *repositories*. They are computer-assisted information systems about the information processing activities of an enterprise. Other commonly used names are *data dictionary, development database, information resource system, catalogue,* or *meta information system* (Ortner 1999).

The most important features of such a repository are the accuracy and recentness, as well as correctness of the contents made available. In order to keep the information about software systems in use, up to date and in concurrence with the upgrading to new systems, releases and thus in concurrence with software configuration management procedures, we have implemented *autonomous agents as pro-active information sources.*

---

[3] This research has been sponsored by the Dresdner Bank AG, Germany.

In the corporate setting of our project partner, the Dresdner Bank AG, software development follows strict rules. One of these includes the use of *archiving servers* for change management and data preservation. Software agents which are monitoring the activities on these servers may provide a highly suitable means to communicate the most recent changes, releases or patches to the corporate development repository in an autonomous and automatic manner.

Currently, Dresdner Bank uses the commercial configuration management tool, *Continuus CM*™ (Continuus 1998)[4] for company-wide software configuration management and archiving server. Both, software development center and repository are located within the corporate intranet. Information agents are extracting the required information. We have implemented a prototype, which has mainly been used to identify the basic goals, abstractions, and plans involved for agents in this problem domain. Furthermore, it provided insights into the applicability, robustness and reliability of agents in this context. Prototypical implementation has been done using the Java-based PRS-implementation JamAgent (Huber 1998a, Huber 1998b); the prototype served as a basis for the final product which will be described in the following.

The modular development and implementation of used agents enables easy future extensions regarding, for example, the interaction with other types of archives or the extraction of different pieces of information. In the current state of the implementation information gathered by the agents is about *released* software systems. Agents communicate via secure channels using XML-encoded messages. The features of tracing and monitoring agents' activities rely on special action log files, which include among others statements in XML on the interaction of agents with some CCM system.

5.2 Domain Characteristics and System Requirements

Software development inside the Dresdner Bank AG is performed in many development centers all over the world. Each of these centers

[4] *Continuus CM*™ is abbreviated by CCM throughout this section. It has been a corporate decision by Dresdner Bank to use CCM for change management, thus most of their software development within c/s-environments is performed with CCM.

produces and archives software according to well-defined procedures and processes. The bank uses a vast number of legacy software, which has been developed for host platforms, mainly MVS-based mainframe systems, in client/server (c/s) computing environments. During the last few years, at Dresdner Bank the problem of change management has been tackled in such c/s-environments by the commercial CCM suite; archiving CCM servers keep track of changes and releases of software.

The main goal of software agents developed in the REPTIL project is to find the most recent releases in the c/s-environment by means of appropriately interacting with given CCM servers. A general overview of the tasks and the flow of information for the agents in their working environment is shown in Figure 14.

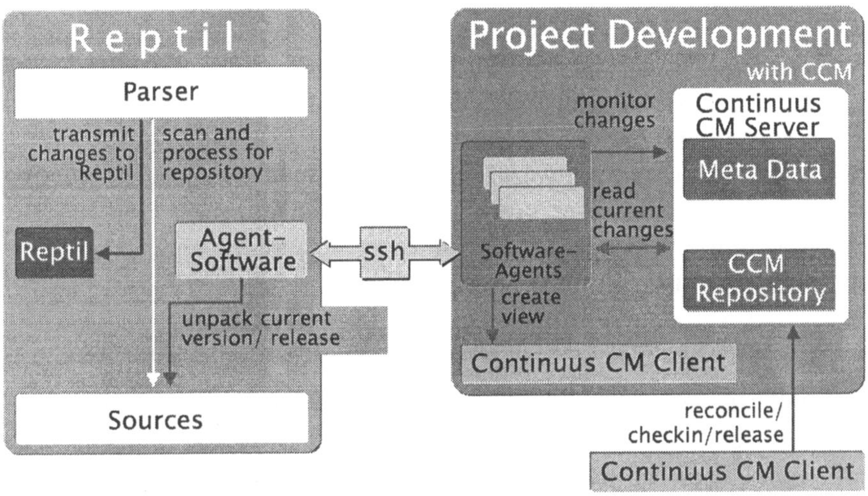

Figure 14. Flow of information between development center (right) and repository REPTIL (left).

This figure also depicts the initial requirements for the agents in general. The project development environment hosts a monitoring agent which interacts with the archiving CCM server, whereas in the so-called REPTIL environment, a software agent receives new incoming information. For obvious reasons, communication and transfer of data including corporate data and values via the network has to be secured.

5.2.1 The Repository REPTIL

The information system, that constitutes the in-house repository solution for the Dresdner Bank AG is called REPTIL and will be commercially available in near-future. REPTIL

- processes, analyses and archives data about software projects and software systems currently in use, and provides respective (meta-) information to the users.
- provides selected operations on archived data for knowledge transfer, information update and refinement, thereby providing a *single point of information* within the software development of the bank.

The *meta information* which REPTIL offers to its users, comprises in particular contextual knowledge about the considered development environment, such as embedded software tools, standards for procedures, lists of users and software components along with an analysis of their data and control flow, as well as organizational entities which are using the software and thus affected by changes to it. In addition, REPTIL provides users with a unified interface in a standard browser (e.g. Netscape or MS Internet Explorer) and is currently applied to the intranet of the Dresdner Bank.

5.2.2 Archive-Based Agent in the Development Environment

We call the special kind of agent in a project development center *archive-based*, since each of the c/s-based CCM server represents one specific kind of *reference location* to be monitored by one or multiple agents. In a more general view of the problem of transmitting the most recent release information as input data to the REPTIL system, such a reference location can be any kind of server or change management and tracking system, like CVS (Fogl 1999), Rational ClearCase™ (Rational), or other sorts of interactive archives[5]. The corporate choice of the bank was the CCM, therefore our reference implementation currently interacts with it.

The agent software located in the development environment has to monitor the contents of the CCM-repository in certain time intervals. It

[5] This does not necessarily have to be a configuration management tool.

looks for specific tags on the CCM-repository's contents. The CCM-repository is organized in terms of (software) projects. Such a project contains all relevant information concerning the continuos evolution of a software system under development. CCM provides a broad amount of operations to keep track of changes and allows for concurrent development including features such as conflict detection. We omit to go into details but refer interested readers to the extensive system documentation (CCM 1998). We limit our description of CCM features to the requirements for the detection of new *project releases,* which is the main task of the agent software.

A new release in the CCM project life cycle is characterized by a *status label*, which has the value *released* (each object under CCM control has a type-specific life cycle; the status label indicates its phase in the life cycle, whereas for projects, the values of this label start with working, and may include *test, integrate* and finally *released* for making the software available). The agent's task is to find those projects, that have been released recently and are not yet updated or known in the REPTIL repository. On detection of a new release, the agent creates its own CCM-specific project view and extracts the respective software and meta data available. The agent's general activity and interaction with CCM comprises the following steps:

- Monitoring of changes at the CCM servers;
- Upon detection of a new, unknown release, the creation of a CCM-specific project view, and
- the collection of the relevant sources code, as well as
- the compilation of available meta information, and finally
- assembling and sending the data packages through the network to the REPTIL system.

These interactions have to be transparent to the software developers using the respective CCM server, to ensure that the agent software does not change substantial system properties.

5.2.3 The Agent in the REPTIL Environment

While the agent software in the CCM environment actively seeks new project information and upon succeeding in this task, gets and sends it, the agent in the REPTIL environment has more simple tasks to achieve.

It waits for new incoming data and upon receipt, unpacks the data and notifies the REPTIL system about it. REPTIL processes the data supplied by the agents, which make up the agent-based information source.

5.3 The Agent-Based Approach

5.3.1 Agents at the REPTIL Site

Since the activity on this side of the agent-based information source comprises only the reception and unpacking of data, the agent-based solution is straight forward.

5.3.2 Agents at the Archiving Server

At the side of archiving CCM servers, we developed an agent society, which consists of a collection of specialized agents (cf. Figure 15). Information extraction from a CCM server is performed by an appropriate task-specific CCM agent. Since there may exist multiple CCM servers to be monitored for relevant data we distinguish between the processes of monitoring the archives, retrieving data from those archives and transmitting data to REPTIL.

- *ADS-Agent:* The *Agent Directory Service Agent* provides several structural services, such as initialization and control of action execution, to the agent society. It observes the activities at the CCM servers. Upon detection of new releases, it invokes a task specific CCM-Agent for information extraction.
- *CCM-Agent:* The *Continuus CM Agent* is activated by the ADS-Agent upon detection of a new project release. It interacts with the respective CCM server for retrieving the source code of and meta-information on a given software development project, compresses the data for transmission to the REPTIL system, and then notifies the RDM-Agent.
- *RDM-Agent:* The *Resource Data Manger Agent* handles the interaction with the agents at the REPTIL site through secure data channels. When the CCM-Agent notifies the RDM-Agent, that data packages are ready for transmission, it informs the agent at the REPTIL site and communicates the data packages through the network.

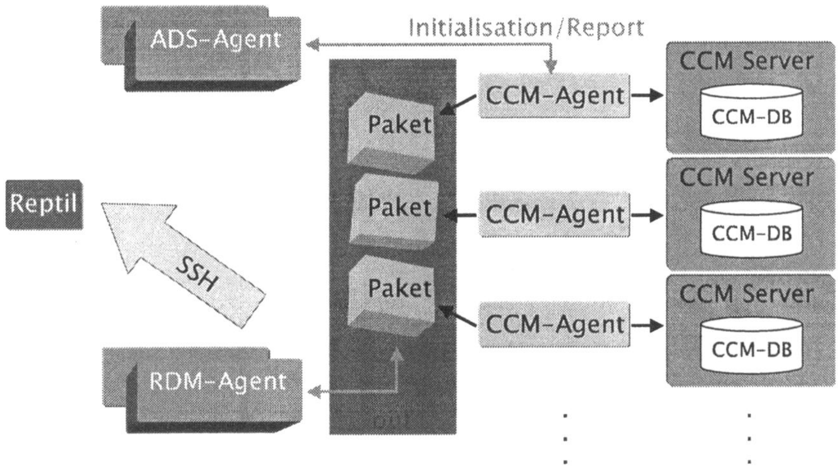

Figure 15. Task-oriented agent society at the archiving server.

Within the REPTIL project we developed a society of task-specific agents, which can easily be adapted to the needs of other archiving servers in the considered software development environment.

5.4 Implementation Details

The agents for REPTIL have been implemented in Java on Sun Sparc machines under Solaris 2.6 using JDK 1.2 and the current JamAgent distribution. We use the Continuus CM™ suite CCM 4.5, script-based interactions are realized in Perl 5. The agents exchange data and message by TCP/IP through secure ssh-channels.

5.5 Future Work on REPTIL

The agent software is currently integrated into the REPTIL system within productive real-life environment of the bank. Potential future enhancements are as follows:

- Interaction with other archiving servers or reference locations, e.g. CVS (Fogl 99) or RationalClearCase™ (Rational).
- Currently the agents and REPTIL provide just information on *released* software. However, CCM also allows for easy switching to different status views (integration test, work, etc.) of projects. This

additional feature enables the repository to present dynamic information about individual software development progress.

Acknowledgments

We'd like to thank Jürgen Feilhauer and Alexander Natter for proof reading and comments on draft versions of Section 5 of this chapter. In addition, we are indebted to Nils Kammenhuber, Patrick Frotscher, Carsten Tichy, and Christoph Stahl for their invaluable efforts in implementing the CASA ITN prototype.

References

Bachem, A. and W. Hochstättler, and M. Malich. Dezember (1992) "Simulated Trading: a New Approach For Solving Vehicle Routing Problems", *Technical Report* 92.125, Mathematisches Institut der Universität zu Köln.

Bürckert, H.-J., Fischer, K., and Vierke, G. (1998), "Transportation Scheduling with Holonic MAS – The TeleTruck Approach," *Proceedings of the Third International Conference on Practical Applications of Intelligent Agents and Multiagents* (PAAM'98).

Bürckert, H.-J., Fischer, K., and Vierke, G. (2000), "Holonic Transport Scheduling With TELETRUCK," *Applied Artificial Intelligence*, vol. 14, pp. 697-725.

Continuus Software Corporation, *Introduction to Continuus/CM™*. 1998, see also http://www.continuus.com

(FIPA) www.fipa.org

(FIPA OS) www.nortelnetworks.com/fipa-os, www.emorphia.com

Fogl, K. (1999), *Open Source Development with CVS.* CoriolisOpen™ Press.

(gACE) http://www.dfki.de/schuhmac/AgentTools

Gerber, A., Klusch, M., Ruß, C., and Zinnikus, I. (2001), "Holonic Agents for the Coordination of Supply Webs," *Proceedings of the International Conference on Autonomous Agents (Agents'2001), ACM Press, New York, USA*

Gerber, C., Siekmann, J., and Vierke, G. (1999), "Flexible Autonomy in Holonic Agent Systems," *Proceedings of the 1999 AAAI Spring Symposium on Agents with Adjustable Autonomy.*

Gerber, C., Siekmann, J., and Vierke, G. (1999), "Holonic Multi-Agent Systems," *DFKI Research Report RR-99-03, ISSN 0946-008X.*

Huber, M. J. (1999), *JAM Agents in a Nutshell, Version 0.61+0.79i,* available at http://members.home.net/marcush/IRS

Huber, M. J. (1999), "JAM: A BDI-theoretic Mobile Agent Architecture," *Proceedings of the Third International Conference on Autonomous Agents (Agents'99),* Seattle, WA, pp. 236-243.

Ingrand, F.F., Chatila, R., Alami, Robert, F. (1996), "PRS: A High Level Supervision and Control Language for Autonomous Mobile Robots", *Proceedings of the IEEE International Conference on Robotics and Automation,* Minneapolis, USA.

Gomber, P., Schmidt, C., and Weinhardt, C. (1998), "Efficiency incentives and computational tractability in the coordination of multi-agent systems", *Proceedings of the Workshop Kooperationsnetze und Elektronische Koordination.*

Kjenstad, D. (1998), "Coordinated Supply Chain Scheduling," *Trondheim, Norway, NTNU-rapport.*

Klusch, M. (1999), *Intelligent Information Agents*, Springer.

Klusch, M. (2001), "Information Agent Technology for the Internet: A Survey," *Data and Knowledge Engineering*, vol. 36, pp. 337-372

Laseter, T.M. (1998), "Balanced Sourcing: Cooperation and Competition in Supplier Relationships," *Jossey-Bass, ISBN: 0787944432.*

Ortner, E. (1999), "Repository Systems," *Informatik Spektrum* 22 (4), pp. 235-251 and 22 (5), pp. 351-363, In German.

Rational.*ClearCase*™ ww.rational.com/products/clearcase/index.jtmpl

Vickrey, W. (1961), "Counterspeculation, Auctions, and Competitive Sealed Tenders," *Journal of Finance* 16, pp 8-37.

Vierke, G., (2000), *TeleTruck - A Holonic Multi-Agent System for Telematics,* PhD Thesis, Saarland University, Saarbrücken, Germany.

Chapter 9

The Use of Virtual Worlds and Animated Personas to Improve Healthcare Knowledge and Self-Care Behavior: The Case of the HEART-SENSE Game

B.G. Silverman, J. Holmes, S. Kimmel, C. Branas, D. Ivins, and R. Weaver

The goal of this project is to determine whether a computer based training game (HEART-SENSE) can improve recognition of heart attack symptoms and shift behavioral issues so as to reduce pre-hospitalization delay in seeking treatment and thereby reduce myocardial infarction mortality and morbidity. In Phase I we created and evaluated a prototype virtual village in which users encounter and help convince synthetic personas to deal appropriately with a variety of heart attack scenarios and delay issues. Innovations made here are: (1) a design for a generic simulator package for promoting health behavior shifts, and (2) algorithms for animated pedagogical agents to reason about how their emotional state ties to patient condition and user progress. Initial results show that users of the game exhibit a significant shift in intention to call 911 and avoid delay, that multi-media versions of the game foster vividness and memory retention as well as a better understanding of both symptoms and of the need to manage time during a heart attack event. Also, results provide insight into areas where emotive pedagogical agents help and hinder user performance. Finally, we conclude with next steps that will help improve the game and the field of pedagogical agents and tools for simulated worlds for healthcare education and promotion.

1 Introduction

1.1 Behavioral and Knowledge Issues in Healthcare

While health outcomes may at first appear to be biologically driven, very few appear to be completely mediated in this way. Virtually all outcomes have a behavioral component at some point in time. For example, many adverse health outcomes can be avoided by simple behavioral change (such as adjusting diet, increasing exercise, or quitting smoking), or by not engaging in high-risk behaviors in the first place. The lay public has a considerable stake in health care and health outcomes, yet many behaviors reveal a lack of either knowledge or concern about these issues on the part of lay persons. For example, while there has been a dramatic decrease in the prevalence of cigarette smoking since the 1966 Surgeon General's report, the incidence of smoking among youths has actually increased, especially over the past five years. In addition, considerable, and at times formidable opposition to anti-smoking legislation (particularly in public places) underscores that a shift in attitude towards a smoke-free environment has evaded some segments of society. Numerous other examples exist, many of which have the weight of legislation behind them, yet there is a failure of most of these efforts to effect real behavioral change that would benefit the larger population. Such issues as motorcycle helmet use, drunk driving, use of appropriate ear or eye protection in certain occupations, and even obtaining minimal prenatal care are examples of how pervasive is the lassitude toward taking effective behavioral measures, by the individual for his or her own safety, improved health status, or prevention of health problems. Not all such programs aimed at reducing risk and improving health status have been failures. The fairly recent programs for educating women about the importance of mammograms in early detection of breast cancer is one stellar success. There are others, but their relative rarity points to a need to assess health education and promotion programs to determine their strength and weaknesses and to develop new strategies.

1.2 Role for Interactive Learning Systems in the National Health Picture

There is increasing evidence that interactive learning systems have an important role in reducing health risks and improving general health status. First, these types of systems have been used extensively in medical student education (Cundick *et al.* 1989, Pincetl *et al.* 1987, Silverman 1992b, Silverman *et al.* 1994) with a great deal of success. In addition, these systems have been deployed for use in continuing education programs (Lajoie and Derry 1993, Psotka *et al.* 1988, Silverman 1992b, Solloway 1993) in such specialties as clinical engineering, military, and language training, to mention a few. Second, numerous investigators have worked on computer-based approaches to patient education, in a variety of clinical areas, such as diabetes (Brown *et al.* 1997, Horan *et al.* 1990, Marrero *et al.* 1989, Meyerhoff *et al.* 1994), heart disease (Liao *et al.* 1996), and asthma (Osman *et al.* 1994, Rubin *et al.* 1986). In a comprehensive literature review and meta-analysis, (Lewis 1999)examined 21 research reports that demonstrated effect sizes (Cohen's D) ranging from 0.21 (mild effect) to 1.91(substantial effect). Most (63 percent) of these studies had effect sizes greater than 0.5 (moderate effect). From this analysis, it is clear that computer-based approaches should be considered as a potential medium for educating patients.

The studies cited above, as well as many others reported in the literature (see Lewis 1999 for a comprehensive review), suggest that computer-based patient education positively impacts on many different dimensions of health and health care, including prevention, compliance, and patient selection of treatment options. While many of these were positioned in interactive environments, none of them investigated the possibility of using game simulation and virtual personas to engage the user.

1.3 Overview of Game Simulators and Virtual Personas

Game simulators have achieved extraordinary popularity in the computer-based amusement market. In addition, "freeware" game simulators proliferate on the web, as well as, assumedly, between and among various peer groups. The reason for their success is simple: they draw the user in, often (but not always) with extensive and sophisticated graphics, realistic (yet fantastic) plots and themes, and engaging interface devices such as sound and emotion. In short, these simulators provide the user with a highly realistic experience that seems somehow "human" in that it contains characters, or *virtual personas*, with whom the user can identify. In fact, virtual personas are ubiquitous in the computing world; they can be found in chat rooms, usenet news groups, and even email. A human user can represent herself in email using a variety of personas simply by using combinations of simple characters in her text (emoticons), to great effect on the person receiving the message. In environments such as chat rooms, the effect is even greater, and is mediated further by the effect of time, since these environments are synchronous, and users can change their personas almost instantaneously.

Game simulators have been used in other settings, as evidenced by the burgeoning educational software market. Products such as the Jump*Start series (Broderbund) and Reading Search (Great Wave Software) have the user go through a series of exercises within the context of some type of adventure game. Without the game, these products would be just so much mundane review of similar concepts. In addition, products such as these usually include one or more virtual personae, often acting as a guide to assist the user. These *animated pedagogical agents* serve to emphasize how much fun the game is to play, cheer on the user when he is doing well, and gently rebuke him when he is performing badly. However, no game-based applications (with or without virtual personas) in health care have been reported, with one notable exception: Alcohol 101. Alcohol 101 (Hardin and Reis 1997, Rhodes *et al.* 1997) is a sophisticated game that is intended to educate the college-aged user about drinking responsibly. The game engages the user by means of two devices: a virtual persona named Norm, and realistic virtual environments such as a party where alcohol is served. The intent of the game is to change (or reify) the user's attitude (and subsequent behaviors) when a choice to drink or not should be made.

1.4 Rationale for Using the Selected Domain

Given the success of Alcohol 101 (it is distributed to incoming freshman by numerous colleges and universities nationwide), a similar approach, using game simulation and a virtual persona, should be plausible in a more clinical setting, especially one in which educating patients and changing their behavior is of key importance. Given the emphasis of the National Heart Attack Alert Program on changing laypersons' behaviors in seeking rapid care for symptoms of acute coronary syndrome, their program appears to be a natural niche for this approach. It is estimated that 26 to 44 percent of the 1.25 million annual heart attack victims delay longer than 4 hours in seeking care for their cardiac symptoms. Yet the greatest reductions of adverse effects occurs if drug therapy is administered within 1 to 2 hours after an attack. Some of the reasons for delay include a bias toward an optimistic outcome (Bett *et al.* 1993); lack of cardiovascular knowledge (Bleeker *et al.* 1995); differences in ethnic background (Chaturvedi *et al.* 1997) and race (Clark *et al.* 1992); prior medical history and non-cardiac co-morbidity (Dracup and Moser 1991, Dracup *et al.* 1997); and denial mechanisms (Esteve *et al.* 1992, Fields 1989).

We hope to reduce delay in seeking care for cardiac symptoms by targeting a decision support game, HEART-SENSE, for use by laypersons as an educational intervention. This system provides a game-based representation of the decisions made by those suffering from a potential myocardial infarction, as well as those who are potential bystanders to such an event (such as friends or family of an at-risk individual). The advantages of this approach over existing methods (beyond those enumerated earlier), such as call centers, written patient education materials, and a printed version of the algorithm are several. First, this approach is self-contained. No other materials, such as pamphlets, guideline sheets, or decision trees are needed. Furthermore, the reasoning behind the decision tree of the game algorithm is hidden from the user. They will not have to use a potentially complicated tree that could be very difficult to follow. Second, the game is instantly available and therefore convenient to use. Regardless of platform, whether palm top or Web, a user can access the game as she would any other software or Web site. Third, written materials such as pamphlets and guidelines, perhaps made available from a local care provider, are sometimes lost or discarded. Furthermore, looking through these materials, even in leisure time, can be time-consuming, and even boring for the user. Placing the message ("rapid identification of cardiac scenarios and appropriate responses") in the context of an entertaining game puts the game within the intellectual reach of a majority of the public, including older children, who may be the only bystander at a heart attack event. Fourth, our approach is cost-effective. The game is free of cost, needing only a personal computer that supports video and sound to run it.

2 Pedagogical Plan

While a game might appear to concentrate on changing short-term behavior among users, its design is grounded in various behavioral and pedagogical theories that point to its likely effect on long-term behavior. Because of this theoretical grounding, we anticipate that field application of the HEART-SENSE game will have a high probability of success, a thesis supported in the results section of this paper. This section explains the theoretical grounding underlying our game.

2.1 Behavioral Theory as Applied to Delay in Seeking Care for Heart Attack Symptoms

Various behavioral theories endeavor to provide a grounding for the development, implementation, and evaluation of programs that are intended to influence personal health practices. These theories have formed the theoretical foundation for sexual health programs, smoking cessation programs, alcoholism, and eating disorder programs. They have also been communicated through a variety of presentation media, including interactive computer software (Hardin and Reis 1997).

The current HEART-SENSE Game is a multi-media interactive, computer-based health promotion program designed to improve health knowledge and influence behavior among individuals thought to be at risk of acute myocardial infarction. In general, any given behavior includes an action (e.g., calling 9-1-1), a target (e.g., a telephone), and a context (e.g., experiencing chest pain). Behavior theorists contend that the most effective interventions will be those directed at a single behavior rather than at multiple behaviors or behavioral categories (Hardin and Reis 1997). In this way, the HEART-SENSE game has targeted "delay in calling 9-1-1" as opposed to a more general goal such as "improved heart health".

In order to effectively change 9-1-1 usage behaviors related to acute myocardial infarction, we have borrowed concepts from several different behavioral models to tailor the HEART-SENSE game to the needs of individuals thought to be at risk of acute myocardial infarction (Ajzen and Fishbein 1989, Bandura 1986, Bandura 1994, Becker 1974, Fishbein 2000, Fishbein *et al.* 1992, Rosenstock *et al.* 1994). This piecemeal strategy has been used in the past (Fishbein 2000, Rhodes *et al.* 1997).

In combining different behavioral theories, care was taken to ensure that the resulting framework was applicable to health behavior problems in general as well as to the public health problem at hand, delay in seeking care for symptoms of myocardial infarction (Fishbein 2000). In preliminary sessions we accomplished this acceptability testing through internal evaluations done by our team of Co-Investigators, both clinical and nonclinical. We followed these internal evaluations with a more detailed assessment of the heuristic sensibility and flow of the HEART-SENSE game by seeking feedback from two panels of clinicians (more on this in Section 5). The advice offered by these panels allowed us to more closely adjust the game and be more confident that our application of behavioral theory was in line with the heuristic knowledge-base among practitioners.

Our integrative health behavior model consisted of three theoretical dimensions that were incorporated into the final version of the HEART-SENSE game. As shown in Figure 1, these three dimensions – intent, skills, and environmental constraints – provided a theoretical grounding for the development of our game (Fishbein 2000). This theoretical basis with which HEART-SENSE has been created will ultimately allow us to confidently proceed to the application phase of the overall project.

Theoretically, a given behavior becomes maximally probable if an individual has a strong intention to execute it, the necessary skills to execute it, and an environment that permits and encourages it. For certain populations, a behavior may not be performed because people have not yet formed the intentions to perform it, while in others, a lack of skills or the presence of environmental hindrances may prevent the performance of a behavior.

The most psychologically complicated of these three dimension is the intention to perform a given behavior. As shown in Figure 1, there are three primary determinants of intention: the attitude toward performing the behavior (i.e. the person's overall opinion of performing the behavior), the perceived norms regarding performance of the behavior (i.e. the perceptions of what others think one should do as well as perceptions of what others would do in the same situation), and the perceived self-efficacy in performing the behavior (i.e. the belief that one can actually perform the behavior when called upon to do so). Our game helps users understand these barriers on a pragmatic plane as we discuss in subsequent sections.

The HEART-SENSE game has also included some statements (made at various times by the village characters and/or the agent coach, Bea) to inform the user of what others think they should do, and what others have done, in a heart attack situation with a positive outcome. In this way, the user is motivated by social pressure that others think they should perform the behavior in question.

The remaining two barriers to behavior change, lack of skills and environmental hindrances, will also play a role in an individual's final decision not to delay when faced with heart attack symptoms. However, both of these barriers are minor relative to an individual's intentions. Usage of a telephone is a common skill that most individuals are able to perform without difficulty. In particular, the 9-1-1 exchange has been promoted by the HEART-SENSE game because of its obvious simplicity relative to seven or 10-digit dialing schemes. Furthermore, the vast majority of the U.S. has been equipped with 9-1-1 technology and individuals have been made aware of its proper uses. Environmental hindrances will also be relatively minor when compared to barriers to intention. Lack of a telephone is probably the most prominent environmental hindrance. Nevertheless, more than 97% of all households in Pennsylvania and more than 99% of households in Philadelphia have telephones making this barrier somewhat negligible (Anon. 2000, Norman 2000).

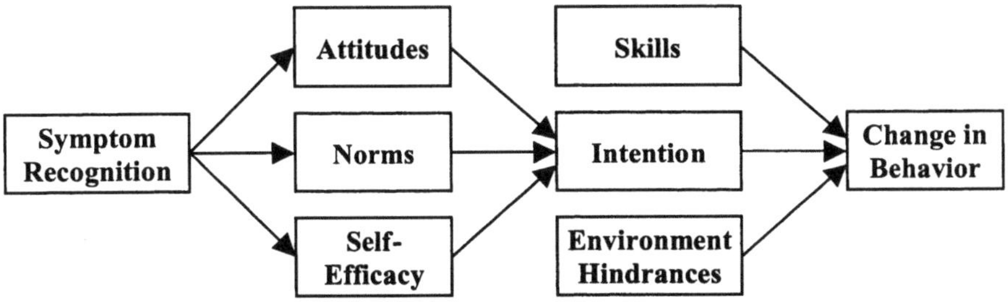

Figure 1. The integrative model of behavior change used with HEART-SENSE. Source: Adapted from (Fishbein 2000).

2.2 Instructionist vs. Constructivist Pedagogy

In the 1980s, prevailing wisdom (e.g., Clancey, 1987) would have driven us to design our environment as an intelligent tutoring system that lectured the users on the behavior change model, and closely drilled them on practice cases. By the end of the1980s, however, many medical schools in North America began to shift from instructivism to constructionism, or problem based learning. At the same time, intelligent tutoring systems fell out of favor in the field of computer science. The idea of constructivist environments caught on, and it is still in widespread use today (e.g., see Solloway 1993).

Evidence shows that the constructivist pedagogy leads to greater internalization and retention of knowledge, increased ability to transfer it to real world situations, and a personal sense of ownership of the knowledge or skill: e.g., see (Elliot *et al.* 1996, Lajoie and Derry 1993, Silverman 1995, Solloway 1993, Wener 1987), among others. Constructivist environments, like problem-based learning, favor student-centered learning, self-discovery of clinical and behavioral insights, and personally experiencing realistic patient cases albeit in a simulated microworld. The more faithful the cases and the microworld is to the real world, the easier the learner can transfer the lessons to actual practice. Also, in constructivist microworlds, if any tutors exist (and often they don't), they act more as facilitators, resources for temporary cognitive apprenticeship, and meta-cognitive aids that focus on learning to learn skills.

Constructivism largely affected how we designed and implemented our interactive learning environment. However, experiments by the senior author indicate that total constructivism may be too unstructured for many learning tasks: e.g., see Silverman (1992a, 1992b, *et al.* 1994, 1995). As a compromise, we add various degrees of instructionism and a personal coaching "agent" to a sound constructivist base as the next section describes.

2.3 The Engage-Instruct-Construct-Persist Training Plan

In particular, Heart Sense consists of three basic modules as Figure 2 depicts, plus a concern for long term follow up. Throughout all modules, an animated pedagogical agent called Bea (heart shaped character) accompanies the user. The "Engage" module includes an attention grabbing sequence that is passively watched by the user. This includes a scenario where a person ignores heart attack symptoms, collapses, and must be rushed by ambulance to the emergency room where it is too late to revive him.

ENGAGE TRAIN REHEARSE PERSIST

Figure 2. Overview of the four modules of Heart Sense game.

This scenario is intended to set the foundation for an individual to adopt a proactive attitude toward performing the necessary behavior (i.e. calling 9-1-1). Thus, if the user can be convinced that performing the behavior in question will lead to "good" outcomes and prevent "bad" outcomes, their attitude toward performing the behavior will become more agreeable (Fishbein 2000).

The emergency room doctor then invites the user into his office for a "didactic" session on how to recognize heart attack symptoms and a range of delay issues. This training tries to use simple concepts and terms that are emphasized by visual aids and animations, plus conversational anecdotes from the Bea character to try and remove the subject's barriers to Intention (attitudes, norms, and self-efficacy). At the end of this "Training" Module, the user is given a short quiz on the critical knowledge. After passing it, the user is asked to go out to the (virtual) community and try to help the residents avoid problems like what just happened to the patient in the Engage Module. At this point the user has been given the basic know-how (reinforced by demonstrating it during the quiz) for helping the virtual residents in the rest of the game.

Next, the users proceed to a village scene where they encounter various residents and hold conversations with them. Some of these simulated characters are in need of help. The pedagogy behind this "Rehearsal" Module is that the user will put their newly acquired know-how to use, help the virtual people to diagnose their symptoms, and convince those same people to change their delaying behavior and seek care without hesitation: i.e., to call 911. This module provides the discovery experience where users explore how delay issues arise, how persistent they are, and how hard it is to convince people to take better care of themselves. The theory is that users will take ownership of this know-how, rehearse it in the safety of a virtual world where their mistakes can be readily "erased," and build skills they can readily translate to the real world for helping both others and themselves. We have also endeavored to bolster the user's self-confidence in actually performing the behavior through positive reinforcement strategies during the game and gentle (not remonstrative) reminders of their errors brokered by BEA.

Thus, in developing HEART-SENSE, we have prominently included attitudinal, normative and self-efficacy elements in order to maximally engineer the intention to avoid delay and call 9-1-1 in the face of myocardial infarction symptoms. Thus, the user's self-efficacy grows stronger as they are shown that they have the necessary skills and abilities to perform this behavior, even in the face of specific barriers or obstacles [11]. The next section of this article explores how the Simulate Module works in more detail.

Throughout the 1st three modules, the user is accompanied by a virtual team-mate, a heart-shaped character named Bea. Bea is an emotive agent with the personality of a cheerful partner who helps users understand module instructions and screen directions, and who encourages and cues the user throughout the various modules. During the simulation module Bea helps by watching the patient condition, game clock, and score, and by alerting the user whenever good or bad events occur. Section 4 describes Bea's architecture, emotions, and logic. Bea is an attempt to introduce a "persona effect", and an appealing (or at least palatable) form of instructionism into the constructivist module.

A final pedagogical concern is that know-how needs to be fresh (near the surface) if people are to be effective in diagnosing symptoms and mitigating delay issues when a heart attack occurs. When users are done with the 1st three modules, they will begin to forget their lessons learned, as the months and years pass by. One way to minimize this problem is to add a module that promotes skill and know-how "persistence". We have not yet constructed this module, but discuss our ideas for it in the conclusions of this article.

3 Interactive Learning Systems and Virtual World

We will begin with an overview of how HeartSense's Rehearsal Module interacts with its users, and then proceed to cover more details. By way of overview, HeartSense works like many other simulators in that it has a case base, a simulation engine, and a user-friendly interface. Working on and solving these cases gives the student practice in symptom diagnosis, behavior alternatives evaluation, and choice. The simulator, in turn, captures the dynamics of the patient's case and the user's evaluations, decisions, and actions as time unfolds. While the simulator primarily serves as a "place" to explore and work out symptom recognition and delay issues, it also coordinates with the Bea agent that provides occasional hints and teaching messages that explain the student's options and decisions.

The next few subsections present each of the components of HeartSense Simulator in turn. The system runs on personal computers under MS Windows. The system is programmed in Macromedia Director, although the case base is an externally editable XML file. The simulator supports statistics collection and student progress and keystroke tracking.

3.1 The Case Base

As discussed above, the case base ultimately will contain multiple people having various types of heart attacks (or not). Each such case consists of 100s of possible dialogue, diagnosis, and intervention options. Each case is described in terms of a state space representation scheme where the patient's situation (one setting of all patient variables) comprises the states, and students' actions determine state transitions. The student controls the simulator clock, so there is ample wall-clock time to explore a given state and alter one's decisions and/or interventions before simulator time advances. To manage the computing complexity, each patient case is indexed into 4 overall levels, including a starting state (L0), a successful end-state and failed-end states (L4), and the remainder being clusters of intermediate states that correspond to moving along the path through the diagnosis (L1), one of the A, N, or E delay levels (L2), and correctly stating the intention to call 9-1-1 (L3). Intermediate states within each level, may be (1) steps in the right direction but from which the student could still branch to either a successful or failed end-state, (2) commonly committed, contra-indicated actions that are important to teach about (and that are recoverable from), or (3) default states that are reached by advancing the clock like “wait and see’, “call the doctor”, or “take and aspirin.” The base of Figure 3 provides an overview of this case base.

To date we have programmed just one full case – the “typical” presentation – and are using that to refine the representations, the simulator engine, and the balance between constructivism and instructionism. Based on the lessons described later in this article, we are planning to improve the design and complete the other cases. To construct even this much of the case base we had to create an XML version of the case file, a Macromedia interface to it, and a web-based authoring environment so case authors could edit the case base remotely from their offices and view the impact on the game (bottom right of Figure 3). This was needed due to the large number of build-test-refine cycles required by our clinical and behavioral team-members to design and deploy even a single case into the case base.

3.2 The Simulator

It is one thing to learn an abstract behavior coping framework when the doctor is lecturing about it (Train Module). It is another thing to know how to apply the framework in realistic settings. The simulator serves as an intermediate step that (a) allows the student to practice and experience what they would ideally encounter in the real world, (b) causes the student to struggle with when and how to structure the multiple recognition and dialoguing dimensions in practice; and (c) helps the instructor to control, broaden, and standardize the learning experience across students, behavior dimensions, and progress to date.

In the simulator, the student goes through the three main modules – engage, train, rehearse. The Engage Module uses audio/video effects in a 2 minute long Director "movie" to portray a scenario where the patient exhibited improper intentions (I) and, by implication, this lead to a bad outcome (e.g., morbidity, mortality). The Train Module reinforces what was just seen but stresses proper behavior, and closes with a quiz intended to reinforce the learning and repetition of the correct answers by the user, and by Bea who coaches them if they err in this quiz. This is another Director "movie" that includes the attending physician giving a 4 or 5 minute slide show to explain the anatomy, symptoms (S), underlying health dynamics, and delay issues (A, N, and E). There are lots of interactive multimedia and animations plus back and forth dialogue between the physician and Bea to keep the material light and entertaining (user affect positive). In programming these two movies (and refining them several times), we have identified some portions of them that can be collected from authors via a template in the authoring environment, though other portions are subject to free and creative expression processes. These are the "teaching materials".

The final module is the simulated "visit", initiated when the user enters the village, selects a village resident, and by that starts a new case. The simulator's algorithm then starts up and cycles through the basic state transitions needed to process the case base and update the current patient dialogs, user response options, and clock and score icons. Each time the virtual patient speaks, the user is given options (see next section) to reply. When one of these is chosen, this triggers the simulator to evaluate the next state, and if it is not a "win" state, then it repeats its basic cycle. If a fail state is reached the user is invited to use the BACK key and create a different outcome (or to exit if they are tired). If a win state is reached, the user is congratulated by BEA and returns to the Village menu, from where they can either help another resident (in a future release) or exit. Each virtual visit takes about 4 to 6 minutes depending on how many states the user visits, though a given case could take significantly longer if the user meanders.

3.3 The Graphical User Interface

Our user interface is designed for use by those with minimal experience in computer software. The only knowledge required is how to move a mouse and click on a response. Even this knowledge will be minimized when the touch-screen kiosk version of the game is completed.

In the Engage Module, the user merely watches the movie. The Train Module requires some minimal interactions such as clicking to start an animated heart, clicking on some buttons that display symptoms or that trigger discussions between the MD and Bea about delay issues, and/or clicking on answer choices for the 4 question quiz. The most complex user interface occurs in the simulator's Rehearse Module, as Figure 3 shows. During the Rehearsal Module when the user encounters a virtual character there is a score bar and time indication in the upper left, an exit arrow on the upper right and forward/back arrows on the bottom right. In each new state of game, once the virtual person and Bea are done speaking, the bottom left shows the responses available to the user. Once the user clicks on one of these choices the simulator branches to the next called for state.

Figure 3. Example screen showing user interface.

4 Animated Personas and Emotive-Cognitive Architecture

One of the strategies for assisting the learning is to include an animated pedagogical agent. The literature suggests this type of agent could make use of artificial life and emotive computing to engage the user, entertain them, and motivate them (e.g., see Lester *et al.* 1997, Silverman 1992b, Silverman *et al.* 1994, Silverman 1995, Solloway 1993) as well as providing a key role to assist and coach them along (Lester *et al.* 1997, Silverman 1995, Silverman 1998, Solloway 1993, Wener 1987). (Isen 1993) shows that trainee performance is significantly altered (statistically so) by tiny interventions such as candy, humor, and the like that induce "positive affect." When affect is positive, subjects learn 50% quicker, arrive at more accurate decisions, and decrease risk-taking behavior. (Lester *et al.* 1997) in turn show that the "persona effect" which is the presence of a lifelike character in an interactive learning environment can have a strong positive effect on students' perception of their learning experience. Silverman (1992a, 1992b, *et al.* 1994, 1995, 1998) shows that situated tutors, coaches, and critics that can assist user judgment during training and performance and can significantly improve know-how learning and transfer. For these and related reasons, we thought it could be worth trying and testing a lifelike character as a "buddy" or partner who goes through the training experience with the user in an effort to keep affect positive, provide a persona effect, and reap the learning and transfer benefits. Understanding how best to incorporate such an intervention and evaluating its value are a goal of our research.

As Figure 4 depicts, such an agent exists as an autonomous process that senses what goes on in the game, decides how best to entertain/help the user within the context of the overall pedagogical plan, and generates and executes the behavior as its effectors on the user learning. The next four subsections describe the building blocks of this architecture. The entire agent including all modules, algorithms, equations, rules, animation stores, and behavior sequences was created by a group of 8 students taking a graduate school course (plus a Teaching/Research Assistant) during the Spring semester of 2000. The agent is programmed in Macromedia Director and the MicroSoft agent server as well as with the help of the H&L Text-to-Speech engine. It is capable of running alongside the game which is also implemented in Director, or independently on the Windows desktop. A web version is a future enhancement.

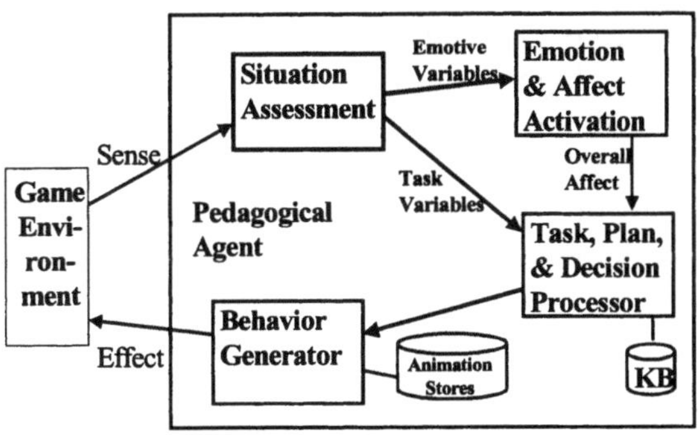

Figure 4. Overview of the Emotive Agent Architecture.

4.1 Affect Theory and Emotive Drives

The basic guidelines for creating animated characters or artificial lifeforms are well-known, although their implementation is subject to some interpretation in any given context. Further, there are a number of unanswered research questions concerning how to ground these lifeforms' emotional drives: (1) so they reflect a logically coherent approach rather than just a designer's whim, (2) so they are not strictly reactive but are more deliberative and purposive with respect to tasks and plans, and (3) so the emotive drives are linked in with cognitive processing. In this section we begin to explore answers to these topics.

In particular, (Lasseter 1987) offers 5 widely used guidelines of cartoon character animation, including: squash and stretch, timing, anticipation, staging, and exaggeration. In addition, several useful guidelines for social interaction are found in (Anon. 1999) such as: create personality, observe appropriate etiquette, use praise, create a team player, consider gender effects, listen-don't joust recognize, clarify and limit choices, provide good error recovery, be efficient and natural, and use appropriate tense, emphasis, and voice. We do not intend to discuss all these guidelines, except to mention that they lead us to the following agent features. Specifically, Heart-Sense uses a character named Bea, who is a heart dressed in a cheerleader outfit and given the personality of a bubbly, energetic, friendly cheerleader who goes through the game as a partner learning along with the trainee. Bea senses what's going on in the game world to form her emotions and then to express them back to the user through her various effectors. Bea's basic effectors are her voice, mouth, eyes, face, and posture and limb motions which are able to represent emotions ranging across ecstatic, happy, calm, nervous, dizzy, worried, and highly alarmed. For example as the right side of Figure 5a shows, when good things happen, Bea can highlight them by cheering and rapid, athletic movements. Or, when the situation is grave, Bea's face and posture reflect it, and her motions slow (Figure 6b). Bea uses readily available lipsynching and voice generation technology. This adds another dimension that enhances her believability and emotional state. Also, when left in idle for too long, Bea stretches, yawns, and takes a nap where her breathing and snoring are visible (but not audible so as to avoid intruding on user problem solving). Finally, Bea's role is primarily to cue or influence proactively, prompt for input, and signal error recovery paths, so her interactions are kept relatively short (most are under 15 seconds), efficient, and non-intrusive.

Various emotive agents in the literature are comparable to Bea in that they react emotively to stimuli from the environment (Anon. 1999, El-Nasr *et al.* 1999, Gratch 1999, Lester *et al.* 1997, Velásquez 1999).

Like many of them, she has an emotional drive module that monitors the environment (game variables in Bea's case) and uses sensed data to activate emotional levels. Bea's basic emotional drives follow the OCC framework (Ortony *et al.* 1988) of reacting to an event or reacting to the actions of another agent (the user), and they are set up as triparites, including an excited, neutral, and inhibited state. For example, a basic emotive drive is "worry over patient condition." When the virtual patient goes from good to fair to poor health, Bea's WORRY state moves from inhibited to neutral to excited. Numerically, these are assigned a 1, 2, and 3, respectively. Aside from patient physical condition, Bea has a JOY drive that decreases from 3 to 1 as the patient's mental condition deteriorates. This is sensed from the game by monitoring a "levels" variable which tracks how far along the patient has moved toward overcoming his/her behavioral obstacles that affect pre-hospitalization delay in the dialog with the user/trainee. There are 6 levels in the game, and this also measures how convincing the user has been thus far. Bea has two other emotional drives as well: one for FEAR of time running out (the game must be played within 30 minutes of elapsed game time – about 10 to 15 minutes of wall-clack time) and DISAPPROVAL for change in plan progress. This last drive measures the relative change in the variable used by the JOY drive (JOY measures absolute change since game start). It reflects whether the user is waivering in convincing the virtual patient to advance along the mental levels.

4.2 Merging Emotions Into Higher Affect to Support Cognition

A concern with the OCC framework is that there is no guidance on when to trigger the different levels of activation of the various emotions. This is left to the agent designer, a state of affairs that is unsatisfying. Also, many emotive characters' emotions tend to be reactions to stimuli, or so-called low-level emotions. Thus fright might lead to flight, or a toe stub leads to pain and anger, but most of us can modulate these impulses. Most of the time, we do not run amok at a loud noise or punch someone who accidentally stepped on our toe. Instead we need to adopt a framework that picks emotive activation levels from principles, and that allows Bea to reason about the emotive signals, factor in knowledge of the broader circumstance, modulate her reactions, and formulate reasonable responses. (Gratch 1999) and (El-Nasr *et al.* 1999) point out that the bulk of existing literature on computational models of emotion focuses on reactive approaches and often have little formal grounding behind the activation dynamics.

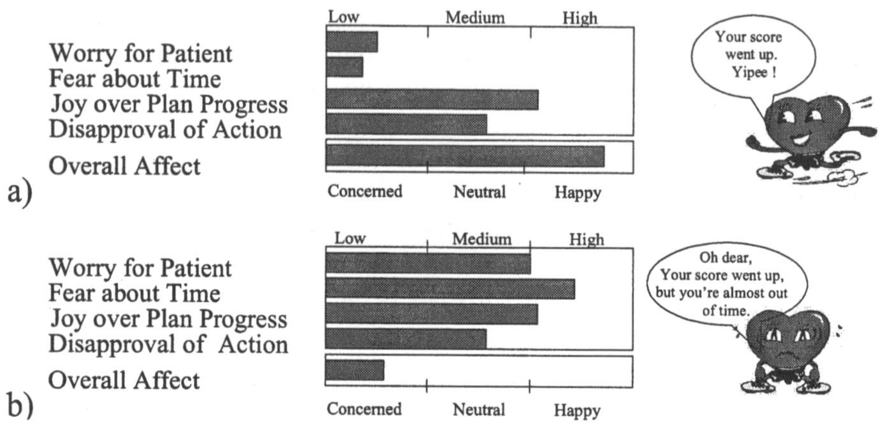

Figure 5. Bea's lower-level reactive emotions are combined into her overall affect to impact how she delivers a given message. Both Figures (a) and (b) are for the same game event (Score goes up due to Plan Progress) but under different Patient Conditions and Elapsed Game Times.

This is a vital problem to overcome if one is trying to utilize emotive agents to foster proactive thinking on the part of the user. The ultimate goal of the emotive agent is to foster skill acquisition and transfer. To help in this effort, the agent must be aware of tasks the user/trainee is attempting and of how well the underlying pedagogical plan is succeeding or failing. For example, the game graph for dialogue between the user and just one virtual patient involves about 80 nodes. A given event may occur during the game (e.g., score goes up from A to B), but where it occurs in the game graph and along the elapsed time scale should alter the emotive reaction of the agent. The agent must have the cognitive ability to make such discriminations.

Our model of agent emotion-to-cognition connection is a generic one that should work for any healthcare game since it factors in virtual patient physical and mental condition, as well as user progress considerations. Ultimately, we wish to derive a single Overall Affect that can be used by the behavior choice module (see next section) to determine which effectors to use in the game. To do this, we first combine the four low-level emotional drive values into two higher level constructs called Efficiency and Effectivity. Efficiency reflects how quickly the user is moving through the game both in terms of elapsed time and progress in moving the user through the levels of behavior change. The mathematics were derived by apportioning time to distance traversal in the game graph, much as industrial engineers would derive efficiency from work progress along a timeline. These equations permit the grounding of the JOY and FEAR emotions, where each is directly correlated to the y- and x-axes, respectively.

In the same fashion, Effectivity combines patient physical condition (WORRY along the y-axis) with user confusion (DISAPPROVAL along the x-axis). If the patient is deteriorated, then the agent should be more concerned about users meandering. Finally, Bea adds Efficiency and Effectivity into Overall Affect (OA). The result gives Bea a deliberative approach to her emotions and overall affect that she then communicates to the Behavior Change Module that we turn to next. This set of equations also explains the left side of earlier Figure 5. There we see an example of how the low level emotions are combined to give different behaviors in 6a vs. 6b even though the same event just occurred for each bargraph (i.e., score rose as plan progress increased).

4.3 Task, Plan, and Decision Processor

The cognitive processes of the agent are to determine what actions (A) to take when a given event or other-agent action, E(x), occurs. Here we view other agent actions as also being events. Since the Bea agent is tasked with helping the user to move efficiently and effectively through the game graph or plan, this means the agent must connect the event to the larger picture and proactively select an appropriate response action. To this end, the agent has a Knowledge Base structured as follows:

$$E(x) \; \& \; OA(H,N,C) \rightarrow A_x(H, N, C)$$

Thus an event E(x) triggers an action, $A_x(H, N, C)$, modulated by whether the Overall Affect is Happy (H), Neutral (N) or Concerned (C). Events may be dialog speech acts, score changes, time changes, and/or patient (physical or mental) condition changes. Actions are names of animation sequences that are parameterized to correspond to the OA(H, N, C) variable setting. Currently it only takes 25 rules to implement Bea's decision processor, though this will undoubtedly grow as more types of events are added to the game, and as we add new responsibilities for Bea as discussed in the conclusions. An example of one of these rules is shown below. The consequent of the rule normally includes only a lookup code, but it's annotated with the text of Bea's behavior to make it readable for the sake of this chapter. In the agent, the text is part of the behavior in the next module. In the Text to Speech engine, an exclamation point is a major inflection and emphasis point, so re-positioning of these punctuation marks can greatly affect the speech act.

IF E(score = 0.5 and score change >0) & OA(.)
THEN:
(HAPPY) A-S1H : "Good job! Your score is going up! You are helping <patient> to recognize that pain is a serious symptom."
(NEUTRAL) A-S1N: "Good. You are heading in the right direction. The pain cannot be ignored."
(CONCERNED) A-S1C : "OK, your score went up by focusing on <patient's> pain. But hurry up! We still have a long way to go."

4.4 Behavior Generator

The behavior generator module is responsible for processing the behavior lookup code (e.g., A-S1H above) and for assembling and executing the proper behavior. This includes retrieving the behavior from the animation store, synchronizing the precise execution interval with the game simulator, and then running the behavior with the help of the Macromedia Director projector and the MicroSoft agent server as well as the H&L Text-to-Speech engine.

The animation store includes about 40 Bea body parts (legs, arms, mouths, eyes, eyebrows, hearts, and miscellaneous items such as motion clouds, water drops, etc.); about 3 dozen low level animation functions written in Macromedia Lingo script (e.g., move, jump, smile, frown, snooze, yawn, etc.); and about 5 dozen higher level behaviors such as the three indicated at the end of the previous section, or the 2 shown in earlier Figure 5. With the help of the Microsoft Agent Editor, the body parts were assembled via a key-frame process into 100s of orientations needed by the low-level functions. Also, via a DirectXtras plugin and a script editor/tester we assembled, the MS Agent environment was merged with the Macromedia Director environment.

Within this editor the various low-level animation functions were assembled and tested. The higher level behaviors were then constructed solely in Director by making numerous calls to the various low level animation functions. Of the 5 dozen high level behaviors, most are on the order of around 15 seconds in length, but a few are explanatory and last for up to 100 seconds. Since Bea's role is intended to assist, tutor, critique in a friendly way, and facilitate, we wanted to keep her mostly in the role of "accent", cuing, and signaling.

5 Usage Results

In the long run we hope to conduct a clinical trial of the full game's impact on personal healthcare intentions, knowledge, and attitude shifts. However, at this point we were interested in evaluating the robust prototype to extract lessons to guide the final design steps. In particular, we wanted to design a "pretest" experiment to: (1) assess usability (e.g., learnability, navigability, clarity), (2) examine pedagogical goal achievement (e.g., within module understandability, across module order, symptom and delay coverage), (3) appraise media alternatives (e.g., impact of audio/video vs. just text on user interest and lesson absorption, value of vivid and animated depictions on memory and discernment, and relative amount of agent emotivity needed to achieve success as a coach and companion), and (4) determine what is measurable and how best to measure it. The results from this experiment will be used to refine the design and help develop the clinical trial concept.

The pretest involved two types of evaluation activities: advisory groups and user tests. The advisory groups included two focus groups and a remote evaluation. The two focus groups were medical faculty and clinicians at the University of Pennsylvania Health System with one group comprising 6 cardiologists and the other group consisting of 6 general practitioners (5 family practice docs, 1 diabetologist). The two focus groups met for 90 minutes each, during which time an attempt was made to guide discussion across a series of 10 topics after a walk-through of the game. The 10 topics straddled the first three items listed above from the viewpoint of how to design the game to best help users. In the case of the cardiologists, a little extra emphasis was given to discussion of whether we had included the right symptoms and of how the game might be made more useful to general practitioners. In the general practitioner session, a little more thrust was placed on whether they felt their patients should use the game and how to make it more useful for that end. In addition, we solicited a scientific review from the American Heart Association (AHA) to help answer the question of whether the coverage of symptoms and delay issues was accurate and pedagogically useful, and whether AHA would host the finished version of the game on their website. For this purpose, AHA was sent a copy of the game to evaluate on their own. None of the focus groups, nor the AHA, were actually shown the list of discussion questions. In the focus groups, the lead author, acting as meeting facilitator assisted by one of the other authors, tried to steer discussion along the lines of the pre-determined 10 questions as best as possible.

The user pretest in turn, was far more structured and consisted of four experimental groups, each seeing a different version of the game as summarized in the four columns of Table 1 and as now will be explained more fully. Version Four of the game consisted of a purely text-based version of the training module implemented as Powerpoint slides in which there were no audio or video media and no animations. Further, there was no back and forth dialog between the MD and Bea, only the doctor's didactic material was included. So this is a didactic version of the game. The other three versions of the game each included all three modules (engage, train, and rehearse), however, they differed from each other on the degree and types of media included. Version Three used the same training module as Version Four, and its Engage Module was stripped of audio and also turned into a power point slide show. Its Rehearsal Module was fully animated with audio and pictures, however, this module eliminated Bea, but had her dialogs popup as small text boxes on the lower right of the screen. Bea's dialog boxes also appeared in the same spot on the power point slides of the Engage and Training Modules for this version. Version 2 and 1 used Macromedia Director for all three modules and permit all forms of media to appear (audio, video, images, animations, back and forth dialogs between virtual personas, etc.). The only difference between Versions One and Two lie in the role for Bea in the Rehearsal Module. There, Version One permits Bea to be fully emotive, while Version Two eliminates her emotions (fear, worry, joy, disapproval). Thus she only appears during the Rehearsal Module of Version Two at the intro screen to explain all the buttons and dials, and at any fail or succeed end states.

Table 1. Overview of the user pretest showing subject demographics and subjects assignments to measurement and to each of four versions of the game.

	Alternative Game Versions			
	4. Power Point (Training Module)	All 3 Modules (Engage–Train–Rehearse)		
		3. Pop-up Text	2. Non-Emotive Agent	1. Emotive Agent
Media Permitted in Version	Text only	Text & pictures in Engage; Text only in Training; All media in Rehearse; Bea dialog as popup text	All media allowed (audio, video, back & forth dia-log, animation); No Bea emotions	Same as Version 3 except fully emotive Bea
Users Assigned to Versions:	4	4	4	4
Age of subjects	1S>55 yrs old 36<2S<55 yrs 1S < 36 yrs	36 < 4S < 55	4S > 55 yrs old 36 < 2S < 55 yrs	2S>55 yrs old 36<2S<55 yrs
Gender (Male, Female)	2F, 2M	1F, 3M	3F, 3M	3F, 1M
Race (C=cauca-sian, M=minority)	3C, 1M	2C, 2M	3C, 3M	3C, 1M
Work (Layman, Health)	2L, 2H	2L, 2H	6L	4L
Heart Disease (None, History, At Risk, Diseased)	2N, 1H, 1D	1N, 2H, 1R	5N, 1H	3N, 1H
Measurements Taken:				
Intentions Quiz Before After	✓ ✓	✓ ✓	✓ ✓	✓ ✓
Knowledge Quiz Before After	✓ ✓	✓ ✓	✓ ✓ (2 S)	✓ ✓
Questionnaire Engage Module (5 Qs) Train Module (7 Qs) Rehearse (11 Qs) BEA Agent (4 Qs) Overall (1 Q) Comments	✓ (3 Qs) ✓ (3 Qs) ✓ (3 Qs) --- ✓ ✓	✓ ✓ ✓ --- ✓ ✓	✓ ✓ ✓ ✓ ✓ ✓	✓ ✓ ✓ ✓ ✓ ✓
Keystrokes Collected	---	✓	✓ (2 S)	✓

Users were assigned to the four versions via an attempt to keep a relatively even distribution across versions for subject age, gender, race, profession, and health status. For a variety of reasons in the pretest (i.e., smallness of sample, date and location in which a version was tested, shortness of time and budget, etc.), distributional randomness was not always possible and some versions got used by subject pools that somewhat favored one demographic over another. For example, Version 4 got used by a doctor and an allied health worker in addition to two laypersons, while Versions One and Two had only laypersons as users. Or subjects using Versions 2 and 3 were 50% minority (african american, hispanic, an/or asian), while only 25% minority was achieved in the groups using other versions. Where these differences may have an impact on results, we will discuss individual responses in the next section.

Finally, Table 1 also summarizes the plan for assigning subject groups to measurement instruments. We collect up to four sets of metrics on each version/group. The fundamental purpose of our research and of our game is to cause a shift in user healthcare behavior and intention. To that end, we pose a before and after Intent Quiz consisting of four scenario questions shown at the top of Table 2. The answers to these questions are open-ended, however, we translate the subject answers to a numerical equivalent score to ease analysis and interpretation. The scoring is as follows:

0 – ignore the symptoms or perform self-care with over the counter medications
1 – wait for a period (often a day or two) and then call the doctor's office or 911
2 – call or go to the doctor's office (possibly after a fairly short wait – an hour or so)
3 – call 911 immediately

Immediately after the first application of the Intent Quiz, we also give a Knowledge Quiz which tests what the subjects know about heart attack symptoms and delay issues. This 2nd instrument is a short 4 question quiz (see Table 2 for exact questions and correct answers) that is also applied immediately after the training module. In Versions 1 and 2, the second application of the Knowledge Quiz is automated with Bea providing feedback and explanation whenever wrong answers are selected. In these versions, the answers are collected in the keystrokes file.

Table 2. Measurement instruments utilized by test subjects.

Intent Quiz

I1. If you had indigestion or heartburn not relieved by antacids, what would you do?
I2. If you had a dull ache in your shoulder, not relieved by pain-killer (e.g., aspirin or advil) and not associated with physical exertion or arthritis, what would you do?
I3. If you had only a mild discomfort in your chest, what would you do?
I4. If you suddenly broke out in a sweat in addition to the conditions in #3 above, what would you do?

Knowledge Quiz

K1. A person having a heart attack always has severe pain in the chest.
K2. Indigestion and heartburn might be the only symptom of someone's heart attack?
K3. If you have symptoms of a heart attack, you should:
  a. Wait for an hour to see if they go away.
  b. Call a friend or relative to get his or her opinion.
  c. Go directly to your doctor's office.
  d. Call 9-1-1 immediately.
K4. Once people know they're having a heart attack they tend to:
  a. Always call 9-1-1 or go to the ER right away.
  b. Call their doctor and wait till he/she calls back.
  c. Delay due to attitudes, social norms, or assumed barriers.
  d. Answers b or c but rarely a.

Questionnaire

Opening Scenario
O1. The instructions are clear on how to use the game.
O2. It is easy to get around the screens.
O3. The opening scenario is clear about the game's purpose.
O4. Photos and audio are of good quality.
O5. The subject matter was depicted vividly and that will help me remember it.

Training
T1. This section (Dr. Levy's Office) is understandable.
T2. The screens and buttons were easy to get around.
T3. I learned something about heart attack symptoms from this.
T4. The explanation helped me see precisely what can go wrong in the heart.
T5. I learned something about pre-hospital delay issues from this.
T6. The material made me think of a time when someone in my experience inappropriately delayed seeking medical care.
T7. It is helpful to have this section before going into the village.

Village
V1. The village scene is understandable.
V2. The instructions for this section were clear.
V3. The screens with Mrs. T were easy to use.
V4. The dialog with Mrs. T was believable and realistic.
V5. I was able to say the things I wanted Mrs. T to hear.
V6. This material helped me better understand how to help people cope with delay issues.
V7. Talking with Mrs. T added to my understanding of what to do during a heart attack.
V8. I am now keenly aware of how quickly time slips by during a heart attack event.
V9. I would relate better to Mrs. T if she were my gender.
V10. I would relate better to Mrs. T if she were my age.
V11. I would relate better to Mrs. T if she were my race.

Table 2 (continued).

Questionnaire (continued)
Bea Character
B1. The Bea character was clear.
B2. The Bea character was believable.
B3. The Bea character was helpful.
B4. The Bea character was encouraging.
Overall
O1. I would recommend this game to people at risk of a heart attack.
Keystroke Metrics
Total clicks
Errorous clicks
Back arrow
Total nodes
Failed ends
Total game time
Total wall time

The 3rd instrument is currently paper-based, and administered at the end of the game before the re-application of any quizzes. This is a 28-item questionnaire that collects user reaction by game module to usability, pedagogical value, logic/believability of characters, and recommendations and suggestions. The precise questions asked are shown in Table 2. All questions are answered on a 7 point Likert Scale with 1 labeled "strongly disagree", 4 labeled "neutral" and 7 labeled "strongly agree". There is also space for free text comments after each module's questions and subjects may select NA for any given question. Since they only saw one module the questionnaire for Version 4 was shortened as will be discussed below.

The final set of metrics collected are keystrokes during the Rehearsal Module. This captures nodes seen, buttons clicked, and time stamps. The results are summarized into 7 metrics (see end of Table 2).

6 Evaluation of Results and Next Steps

The results from the advisory group sessions and from applying these various instruments and measures may best be summarized in terms of six topics – intention shift, pedagogical value, usability, logic/believability, persona effect, and miscellaneous. In terms of intention shift, the results from the Intent Quiz indicate that users in all four groups experienced a positive shift over all four intention categories, as shown in Table 3a. Although the sample size (N=18) is too small to assess the statistical significance of these results, one can see trends in the effect of the various levels of intervention. The subjects exposed to the full version of Heart Sense and those exposed only text-based materials experienced the highest shift (125.0 percent and 140.6 percent, respectively), while those exposed to the mixed text/non-emotive agent version had the lowest shift (37.5 percent). In terms of specific dimensions of intention, subjects tended to move away from the intention to care for themselves or to wait to call their doctor in the event of a heart attack, after being exposed to either of the two multimedia versions of the intervention, more than those who were exposed to the other two versions. All versions of the intervention, except for Version 3, effected a shift in intention to call 911. While the shift is most pronounced in the text-only version, it is prominent in the two computerized interventions.

Table 3a. Mean Intention shift (as percent change between pre- and post-intervention) for 4 dimensions across the 4 versions. Numbers in parentheses represent one standard deviation.

Intention	Version 1 Full Emotive	Version 2 Basic Bea	Version 3 Popup Text	Version 4 Didactic/Text	TOTAL
• Self care	100 (115.5)	166.7 (116.9)	37.5 (47.9)	75 (150)	102.8 (115.7)
• Wait to call	112.5 (165.2)	158.3 (128.1)	87.5 (85.4)	200 (141.4)	141.7 (127.5)
• Call/see doctor	200 (141.4)	83.3 (132.1)	37.5 (47.9)	125 (150)	108.3 (128.6)
• Call 911	87.5 (143.6)	50.0 (77.5)	−12.5 (25.0)	162.5 (75)	69.4 (101.6)
TOTAL	125 (141.4)	114.6 (113.7)	37.5 (51.5)	140.6 (129.1)	105.6 (118.4)

Grouping together the two computerized versions, the difference in effect on intention is even more striking (Table 3b), where the shift was 118.8 percent in the former, and only 89 percent in the latter. This indicates that any multimedia version of the intervention (emotive or non-emotive) results in a positive shift in intention. This was especially evident in the intent to continue to care for one's self in the event of a heart attack, where the difference was 140 percent vs. 56.3 percent for the computerized and non-computerized versions, respectively. Although the shift intention to call 911 was essentially the same between the two groups, the intention to call or see a doctor was substantially positively shifted in the multi-media versions. The important finding in this analysis is that there is a positive shift in intention apparently caused by the full multi-media and mixed pedagogy versions versus the didactic, text-only approaches taken by Versions 3 and 4.

Table 3b. Mean Intention shift (as percent change between pre- and post-intervention) for four dimensions across the multimedia vs. the didactic versions. Numbers in parentheses represent one standard deviation.

Intention	2 Animated Versions	2 Textual Versions
• Self care	140 (115.0)	56.3 (105.0)
• Wait to call	140 (137.0)	143.8 (123.7)
• Call/see doctor	130 (141.8)	81.3 (113.2)
• Call 911	65 (102.9)	75 (106.9)
TOTAL	118.8 (124.2)	89.1 (112.2)

These intention shifting results were predicted by the pedagogical theory the game is grounded in (Section 2). They were also predicted by AHA's Scientific Review Panel and that explains why they approved Heart Sense without even a single revision required. For the same reasons, after seeing Version 1 of the game, both focus groups of physicians devoted a sizable fraction of their total time to discussing their concern that if this game were rolled out nationally and widely played that the ERs could not support the influx of added visits. Both groups strongly favored such a roll out, and neither group was willing to water down the message, however. They favored sensitivity errors (false positives) to specificity ones (false negatives) to the extent of insisting that no false positive cases should be included in the village – all villagers should have true MIs.

Looking at the Knowledge Quiz overall impact, all versions effected a change in knowledge (Table 4a), although the didactic-only version (V4) did the best at improving after-the-fact learning The group using that version is the only group with no after-game errors on the quiz. All other groups had 19 to 25% of their answers with after-training errors.

One interpretation of the improvement in Version 4 might be that the other versions did not teach the answers as succinctly, given that they introduced distractions via the Engage Module, the plot involving the various personas, and various audio/visual effects, and so on. Indeed, one of the physicians in the Practitioner Focus Group pointed out that this was the case, and more importantly stated that the quiz questions did not fully reflect what was emphasized in the full multi-media versions of the Engage and Training modules. Another interpretation might be that Group 4 had 50% health workers (although Group 3 did as well) as opposed to all laypersons in most of the other groups. Also, when we look at individual questions, we see that Group 4 had the least number of wrong answers when initially taking this quiz – so maybe this was not a representative group in terms of prior knowledge.

Table 4a. Mean Knowledge Quiz scores obtained before and after intervention for each of the four intervention versions. Scores ranged between 0 and 4. Numbers in parentheses represent one standard deviati on.

	Pre-Intervention	**Post-Intervention**
Version 1 – Fully Emotive	2.75 (0.96)	3.33 (0.58)
Version 2 – Basic Bea	1.83 (0.75)	3.0 (3.05)
Version 3 – Popup Text	2.75 (0.96)	3.25 (1.5)
Version 4 – Didactic/Text	3.0 (0.82)	4.0 (0.0)
TOTAL	2.5 (0.92)	3.5 (0.85)

However, in comparing the changes in knowledge score between the multi-media and the text-based versions, there is a substantial difference between those subjects exposed to these two interventions. Table 4b demonstrates that the multi-media versions resulted in a substantially greater improvement in knowledge, post-intervention.

Table 4b. Percent difference between pre- and post-intervention Knowledge Quiz scores for each of the four intervention versions.

	Percent difference in Knowledge Quiz score
Animated/Multimedia versions	51.4%
Didactic versions	25.1%

The Questionnaire was designed to provide insight into the user's reactions to various aspects of the intervention. Table 5 shows the raw results of the Questionnaire data for each subject.

While these data are useful in their raw form, they are perhaps most interpretable when grouped into three dimensions that reflect the main metric foci of these data: *Usability*, *Logic*, and *Pedagogy*. Table 6 shows the Questionnaire scores grouped by these dimensions:

Table 5. Raw results form the Heart Sense Questionnaire.
1 = Highly disagree with statement or question; 4 = Neutral;
7 = Highly agree; NA = Not applicable to this version.

Version:		Emotive Bea				Non-Emotive Bea						Pop-up Text				Power Point			
subj. no.:		1	2	3	4	1	2	3	4	5	6	1	2	3	4	1	2	3	4
Opening Sc.	O1	7	7	6	7	7	6	7	7	4	7	5	7	6	7	5	6	7	7
	O2	7	7	7	7	6	7	7	7	6	7	5	7	7	7	7	7	7	7
	O3	7	7	7	7	7	7	7	7	6	7	5	7	7	7				
	O4	7	7	7	7	7	6	7	7	6	7	6	7	7	7				
	O5	7	7	7	7	5	7	7	7	6	7	6	7	6	6	5	5	7	7
Training Section	T1	7	7	7	7	7	6	6	7	6	7	5	7	7	6	6	6	6	7
	T2	7	7	7	7	7	6	7	7	6	7	5	7	6	7				
	T3	7	7	7	7	5	7	NA	7	6	7	6	7	7	7	5	5	7	7
	T4	7	7	6	7	6	7	7	7	5	7	4	7	6	7	6	6	3	7
	T5	7	7	7	7	NA	7	7	7	5	6	6	7	7	7	6	6	5	7
	T6	5	7	7	1	7	NA	NA	NA	6	6	6	NA	7	7	4	4	NA	6
	T7	7	7	6	7	7	6	4	7	6	7	6	7	7	6				
Village Section	V1	7	7	7	7	7	7	7	7	6	7	5	7	7	6				
	V2	7	7	7	7	7	6	7	7	4	7	6	7	7	6				
	V3	7	7	7	7	7	7	7	7	5	7	6	7	7	7				
	V4	7	6	7	5	7	6	6	7	5	7	6	7	7	7				
	V5	7	6	7	7	7	6	7	7	5	7	6	7	7	5				
	V6	7	7	7	7	7	7	7	7	6	7	6	7	7	7	5	6	5	7
	V7	7	7	7	7	7	7	7	7	5	7	6	7	7	6				
	V8	7	7	7	7	7	6	7	7	5	7	6	7	7	7	5	5	6	7
	V9	1	NA	NA	1	4	3	NA	NA	3	NA	4	1	4	1				
	V10	1	1	4	1	4	3	1	2	3	NA	4	1	4	1				
	V11	1	1	4	1	4	NA	NA	2	3	NA	4	1	4	1				
Bea Character	B1	6	6	7	7	7	7	7	7	6	7	NA	NA	NA	NA				
	B2	5	4	7	7	3	2	NA	7	6	6	NA	NA	NA	NA				
	B3	7	6	7	7	7	6	7	7	6	7	NA	NA	NA	NA				
	B4	6	4	7	7	7	5	7	7	6	7	NA	NA	NA	NA				
								(a)											
Overall	O1	6	7	6	7	7	6	7	7	6	7	7	7	7	7	7	6	7	7
				(b)								(c)		(d)				(e)	
Demographics		3	4	3	4	4	4	4	4	3	3	3	3	3	3	3	4	2	3
		F	F	F	M	M	M	F	F	M	F	F	M	M	M	F	M	F	M
		C	AA	C	C	AA	C	C	H	C	AA	C	A	C	A	C	C	C	AA
		none	none	none	none	none	none	none	none	none	none	allied	none	limited	doctor	allied	none	none	none
		N	N	N	H	H	N	N	N	N	N	HH	R	H	N	HN	D	H	N

Comments:

(a) Bea was cute

(b) could be a little shorter

(c) Should use to train general population – e.g. college and HS students

(d) I think this software is very useful because it is informative about the sources and risks of delays in seeking appropriate heart attack treatment

(e) wording was awkward on some slides which slowed me down

Table 6. Mean questionnaire scores grouped on three dimensions across the four intervention versions. Numbers in parentheses represent one standard deviation.

	Version 1 Emotive	Version 2 Basic	Version 3 Text	Version 4 Didactic
Usability	48.8 (0.5)	46 (4.6)	44.8 (5.3)	19.5 (1.3)
Logic	25.0 (3.5)	22.8 (3.8)	24.5 (3.1)	3.75 (2.6)
Pedagogy	55.5 (1.0)	49.8 (5.7)	52.3 (4.3)	34.8 (4.9)

The *usability* dimension is vital since many of our users may not be very computer savvy. This dimension was addressed by the focus groups and by users responding to items in each section of the questionnaire including O1, O2, T1, T2, V1, V2, V3, and B1. The focus groups complained that the score and clock icons in the upper left of the Rehearse Module need refinement. They felt time is a particularly important variable for users to learn to manage and voted for a continuous analog rather than discrete digital clock thinking that might make it more intuitive. Also, they wanted it to count down the time remaining, rather than increasing as the score does. This would help to differentiate it functionally from score bar. Aside from these concerns, all user reactions place the usability of all versions in the positive realm. However, the strength of usability ratings are highest for Version 1 and drop a little further for each successively media-crippled version. It is interesting to note that Version 4 has the lowest ratings despite being the shortest module with the least complexity (touch any key for the screens to advance in the standard PowerPoint slide show fashion). This attests to the success of our graphical user interface, the use of simple mouse actions, the role for Bea, and the impact of multimedia. In fact, this is borne out in Table 6, which demonstrates that the three computerized versions substantially outperform the text-only didactic version. In addition , a trend can be seen in which versions with decreasing levels of multimedia and emotive interface interaction correlate with decreasing usability. Although the number of subjects examined in this pilot are far too small to make any statistical inferences, this trend supports the three points made above.

The Questionnaire was also designed to provide insight into the detailed impact of one pedagogy versus another. See Table 5 for a detailed description of the results of the Questionnaire data pertinent to the pedagogical dimension of the intervention. Versions 1 to 3 (multi-pedagogy approaches) result in:

(1) vividness that helps them better remember the results (O5);
(2) learning about symptoms overall (T3), precisely where symptoms occur in the heart (T4), and about delay issues (T5, V6) – which may help to explain the observation that Version 4 subjects had less to learn prior to the intervention;
(3) subjects learned more keenly how time slips by during a heart attack event (V8).

These are precisely the areas where one would expect the multi-media and simulation/rehearsal effects to have an impact relative to the pure text-only didactics of Version 4. Again, the data in Table 6 show clearly the effect of the three computerized versions on the pedagogical dimension, with the fully multimedia, emotive-agent version having the strongest effect. The slight decrease in score for Version 2 on this dimension may be a function of sample size, or it could reflect the "unfriendliness" of the interface for this version, relative to those of Versions 1 and 3. Regardless, there is evidence of a trend of decreasing pedagogical satisfaction that correlates with decreasing interface detail.

The final Questionnaire dimension allows us to probe is how *logical* (and believable) are the virtual characters and their scenes. We placed at least one question about this in each section of the questionnaire and coverage is provided by O3, T6, V4, and B2 in Table 5. We can see in Table 5 that virtually all subjects strongly or very strongly feel the characters and their situations are believable and logical. One exception to this appears in Subject 4 of Group 1, who was apparently unmoved by Bea's attempt in the Training Module to cause him to reflect on people in his past who had delayed in seeking health care. All other users were stimulated to do so, even (though less so) the users in the didactic Version 4 who did not benefit from Bea's spoken anecdotes. In general the least strongly believable character was Bea as answers to B2 indicate. Here there was a somewhat bipolar split where subjects in each group felt either strong/very strong or they felt less so. The non-emotive Bea drew worse believability scores than the emotive Bea. Given that Bea is a cartoon character, and all the other characters are photographs of actual actors, it is not surprising that Bea is the least believable. It might even be surprising that she is believable at all, and most users probably should have answered this question as NA (only 1 answer was NA). However, another interpretation may be what was voiced by one of the physicians in the focus group. He felt that Bea was too juvenile as a persona and that she didn't connect with him. He viewed her as being of high school, and more appropriate for helping kids play the game rather than for middle-aged persons and senior citizens. This view was not shared by any others in the focus groups, but it may be that the two males who rated her 2 and 3 in Group 1 had similar reactions. In subsequent research we hope to examine reactions of users to alternative voices and age levels of the Bea character. The major collector of emotive toys in America is turning out to be middle-aged women, a target group we particularly want to reach with our game. In fact, the only free text comment we got from any user in this portion of the questionnaire was from an over 55 year old female who was moved to comment that "Bea is cute" (the same individual who appropriately stated NA on the believability question). So it is not a foregone conclusion how we should handle this issue.

Related to the believability issue are items V9, V10, and V11 in the questionnaire which explore whether the user would relate better to the character having the heart attack if that character were the same age, gender and race as the user. Almost two thirds of users were negative about this, and another 29% were neutral. While no one was in favor of this, over a third felt it was not applicable for them to answer one or more of these three questions (possibly since Mrs. T already matched part of their profile). The same physician who felt Bea was too young, also felt strongly that Mrs. T was too old and the wrong gender. If our subject pool had involved less professional people, it is possible we would have encountered more of the type of reaction this physician claims his patients would give. It is fairly straightforward (though tedious) to alter the case base of the simulator so that it can be age/gender/race responsive. A somewhat more difficult task is to determine which age/gender/race the user prefers to see in the village. Some of these demographics are not so easy to determine as subjects may not wish to answer about them, or as some subjects might be playing for someone else they provide care for. We plan to explore this more fully in subsequent research, and it might also be interesting to explore dynamic personalities for the Bea character as well.

Table 6 shows the apparently strong effect of computerization on the perception by users that the material is logical. The difference between the scores on this dimension between Versions 1 through 3 and Version 4 is very striking, and causes one to wonder if the text-based materials are at all effective in bringing the reader (user) "in" to make the pedagogical experience more meaningful. Clearly, one must question the efficacy of text-based materials, specifically in this study, and perhaps in general, when results such as these are found.

The field of human computer interaction is split over the value of a character such as Bea for coaching and assisting the user. Some researchers argue that such an agent is distracting, hampers users' behavior, and potentially reduces text retention (Dehn and Mulken 2000). Other researchers argue that "animated pedagogical agents" play a critical role in helping and motivating students, the more so the more they display realistic emotions (Lester *et al.* 1997, Elliot *et al.* 1996). Our results lend partial support to each view and shed some light on where animated pedagogical agents are needed and where certain behaviors are inappropriate.

Let us begin by pointing out that our entire game (all 3 modules) consists only of animated or virtual characters. In the Engage Module, Bea is added to the mix of other animated characters to provide user companionship, entertainment, and emotional bonding, rather than for purely pedagogical purposes. In the other two modules Bea is a secondary pedagogical agent, although we hadn't realized it ahead of time in the latter of these. In the Training Module the primary pedagogical agent is Dr. Levy who gives the didactics. Bea's role there is to simplify the message, translate it to everyday terms, and keep the presentation conversational and anecdotal. Bea plays the identical role for both versions 1 and 2 in the Engage and Training Modules. We believe this is why Bea was uniformly labeled strong (6) or highly strong (7) on being clear (B1) and helpful (B3) across all users of versions 1 and 2, even by those who were neutral or negative on whether she was believable (B2) or encouraging (B4). In retrospect we regret not having phrased the questionnaire so users of version 3 could have answered B1 through B4 with other than NA for their popup text boxes. However, we suspect their slightly lower game usability ratings are related to the lack of Bea's persona.

In the Rehearsal Module we originally conceived Bea as the pedagogical agent (using cueing, hinting, and muted suggesting). We feared users without Bea would stray off the normative path, incur large delays in the game clock, and potentially reach failed end states. However, as one can see from the Keystroke results (Table 7) users of Versions 1, 2, and 3 did equally well, minimal straying off the correct nodes occurred, everyone finished well within the allotted 30 minutes on the game clock (4 to 6 minutes wall clock), and no failed end nodes were reached. Thus there was no user performance difference (improvement or decrease) due to Bea's emotivity being on/off or even Bea having a physical or audible persona (recall that in version 3 Bea is just popup text boxes). The only significant difference was that the fully emotive Bea caused the Rehearsal Module to last an extra 1 to 1.5 minutes which prompted one user to suggest the game "could be a little shorter". The reason for this persona non-effect is that Mrs. T is herself an animated agent sending strong verbal and visual cues that she needs speedy help (typical presentation with crushing chest pain). We suspect that as less obvious heart attack scenarios are introduced via other village characters, that users may be less clear on what to do and that Bea's pedagogical role will become relevant in this module as it is in the earlier ones.

Table 7. Summary of keystrokes for emotive, non-emotive, and popup text versions.

version_user	total nodes	errorous nodes	back arrow	total correct nodes	failed ends	total game mins.	total wall time (secs.)
v1_1	21	6	0	15	0	20	349
v1_2	19	4	0	15	0	18	347
v1_3	20	5	0	15	0	19	345
v1_4	18	2	0	16	0	17	325
v2_5	21	5	1	15	0	20	255
v2_6	19	4	0	15	0	18	276
v3_1	19	4	0	15	0	18	237
v3_2	18	3	0	15	0	17	276
v3_3	19	4	0	15	0	18	282
v3_4	21	5	0	16	0	20	347

To summarize, a pedagogical agent like Bea seems to be most helpful to reduce complexity, provide cues, offer companionship, and enhance entertainment value to keep user affect positive and promote learning performance. The animated pedagogical agent approach appears to become irrelevant where the situation (or other characters) already provide a sizable portion of these items on their own. Also, we are unable to concur with (Wright *et al.* 1999) that animated pedagogical agents reduce text retention even though our version 4 subjects tested the highest since our Knowledge Quiz (and possibly Group 4's subject pool) was biased away from the material taught via the animation. Finally, our cases to date did not permit us to collect fine grained feedback on when high degrees of emotivity helps or hurts, and this will have to be a topic for our future investigations.

Finally we come to the last set of results to be discussed in this section – the results that include miscellaneous lessons learned. From the focus groups we learned several insights about the Engage Module. For one thing, it is currently too slow moving, actors aren't sufficiently energetic, and has material that needs to be edited and tightened up. A more serious concern is that all physicians objected to the fact that our patient dies in the emergency room. This is not just a complaint that competent doctors would never allow it to happen. Given that we are trying to teach an intention shift concerning calling 911 and going to the ER, this scenario conveys a negative message we don't want our users to infer about ERs. We will be altering this part of the scenario in the next release. Also, in the Training Module the Doctor hands Bea an AHA pamphlet, which she reads from and then shows portions of on the screen. The doctors felt this should either be something the users can printout and take home, or not be shown as a pamphlet. Finally, all focus group doctors and all users felt strongly (6) or very strongly (7) that the game should be shown to people at risk of heart attack (Overall Question 1).

7 Conclusion and Next Steps

7.1 Contributions to Date

Several innovations have occurred during this effort. First, this effort has lead to the design of a generic simulation environment and authoring toolset that helps to implement patient behavior change theory. The design permits game authors to mix instructionism and constructivistic pedagogies, multi-media, and animated pedagogical agents to enhance user learning experiences and performance. Having such an intentionality theory-grounded toolset as a generic capability would be a first for the field of healthcare education.

Also, the design provides a mathematical grounding for agents to dynamically calculate realistic emotional response to changing game states. In the field of emotive agents, the sensing of changed environmental states is commonplace, but choosing which emotion to then activate is often based on ad hoc or experientially derived table style lookup. Mathematically grounding the emotion activation decisions in game planning space makes the agents generic to any healthcare domain, and this eases development effort.

A proof of concept prototype has been created in the heart attack domain that illustrates these various innovations. Initial evaluation studies of that prototype also lead to it passing AHA scientific review, physician focus group predicting that a national rollout of the game would result in Emergency Rooms being filled to capacity and beyond; and statistically significant user shifts in intention to call 911 and avoid delay. User test results also indicate the full multi-media and animated agent versions of the game provide improvement in learner understanding and memory of symptoms, and appreciation of time management during heart attack events.

User reaction data indicate the emotive pedagogical agent improves usability overall as well as learner performance during complex explanations. Further the emotive agents are viewed as highly useful as companions and for entertainment that appears to keep user affect positive. Emotive and animated pedagogical agents appear irrelevant when the situation and/or other characters make it clear what is required of the user. These findings might be described as a corollary to (Lester *et al.* 1997)'s "persona effect" which states that animated pedagogical agents are *always* helpful, credible, and entertaining, but more so the more expressive the agent. Our corollary, which we call "situational dependency", seems to indicate that (1) different types of emotivity and persona are *credible* to different users, and (2) animated pedagogical agents are *helpful* only where needed to boost companionship, increase entertainment value, and reduce complexity. Thus if the situation is already understandable and companionship and entertainment offer little added value or if there is a mismatch between the virtual agent and user's personalities, then there will be no persona effect and one can safely omit the pedagogical agent from that situation.

7.2 Next Steps

What's past is prologue, and the results to date provide a useful roadmap to further research needs for both the Heart Sense Game and for the field of simulators and pedagogical agents for healthcare in general.

In future research we hope to investigate a number of these issues.

Acknowledgments

We gratefully acknowledge the financial support of the National Heart Attack Alert Program of NIH, and the National Library of Medicine PO # 467-MZ-902060. Also appreciated are the advice and support of the scientific review team of the American Heart Association, Marty Fishbein at Penn, and the 29 focus group members and user testers. Also, Yi Chen assisted with programming the simulator.

References

Ajzen, I. and Fishbein, M. (1989), *Understanding Attitudes and Predicting Social Behavior*, Prentice Hall, Englewood Cliffs, New Jersey.

Avis, N.E., Smith, K.W., and McKinlay, J.B. (1989), "Accuracy of perceptions of heart attack risk: what influences perceptions and can they be changed?," *American Journal of Public Health*, pp.79: pp.1608-12.

Bandura, A. (1994), "Social cognitive theory and exercise of control over HIV infection," in Diclemente, R.J. and Peterson, J.L., (Eds.), *Preventing AIDS: Theories and Methods of Behavioral Interventions*, Plenum Press, New York, pp. 25-60.

Bandura, A. (1986), *Social Foundations of Thought and Action: A Social Cognitive Theory*, Prentice Hall, Englewood Cliffs, New Jersey.

Becker, M. (1974), "The health belief model and personal health behavior," *Health Education Monographs*, vol. 2, pp. 324-473.

Bett, N., Aroney, G., and Thompson, P. (1993), "Impact of a national educational campaign to reduce patient delay in possible heart attack," *Aust NZ J Med,* vol. 23, pp. 157-161.

Bickmore, T.W., Cook, L.K., and Churchill, E.F., et al. (1999), "Animated autonomous personal representatives," *Proc. 2nd International Conf. on Autonomous Agents*, ACM Press, New York, pp. 8-15.

Bleeker, J.K., Lamers, L.M., and Leenders, I.M., *et al.* (1995), "Psychological and knowledge factors related to delay of help-seeking by patients with acute myocardial infarction," *Psychotherapy & Psychosomatics*, vol. 63, pp. 151-158.

Brown, S.J., Lieberman, D.A., Germeny, B.A., Fan, Y.C., Wilson, D.M., and Pasta, D.J. (1997), "Educational video game for juvenile diabetes: results of a controlled trial," *Medical Informatics*, vol. 22, no. 1, pp. 77-89.

Carleton, R.A., et al. (1996), "Report of the Expert Panel on Awareness and Behavior Change to the Board of Directors," American Heart Association, www.americanheart.org/Scientific/statements/1996/059601.html.

Chaturvedi, N., Rai, H., and Ben-Shlomo, Y. (1997), "Lay diagnosis and health-care-seeking behaviour for chest pain in south Asians and Europeans," *Lancet*, pp. 350: pp. 1578-1583.

Clark, L.T., Bellam, S.V., Shah, A.H., and Feldman, J.G. (1992), "Analysis of prehospital delay among inner-city patients with symptoms of myocardial infarction: implications for therapeutic intervention," *Journal of the National Medical Association*, pp. 84: pp. 931-937.

Cundick, R. and Turner, C.W., *et al.* (1989), "ILIAD as a patient case simulator to teach medical problem solving," *SCAMC Conf. Proc.*, pp. 902-906.

Dehn, D.M. and Mulken, S. (2000), "The impact of animated interface agents: a review of empirical research." *Int. J. Human-Computer Studies*, vol. 52, pp. 1-22.

Dracup, K., McKinley, S.M., and Moser, D.K. (1997), "Australian patients' delay in response to heart attack symptoms," *Medical Journal of Australia*, vol. 166, pp. 233-236.

Dracup, K., Moser, D.K., and Eisenberg, M., et al. (1995), "Causes of delay in seeking treatment for heart attack symptoms," *Soc. Sci. Med.*, vol. 40, no. 3, pp. 379-392.

Dracup, K. and Moser, D.K. (1991), "Treatment-seeking behavior among those with signs and symptoms of acute myocardial infarction," *Heart & Lung*, vol. 20, pp. 1-5.

Elliot, C., Williams, K., and Woolf, B.P. (1996), "An intelligent learning environment for advanced cardiac life support," *Proc. AMIA Conf.*, Bethesda: AMIA.

El-Nasr, M.S., Ioeger, T.R., and Yen, J. (1999), "A web of emotions," in Velásquez, J. (Ed.), *Emotion Based Agent Architecture Workshop at Autonomous Agents'99 Conference*, www.ai.mit.edu/people/jvelas/ebaa99.

Esteve, L.G., Valdes, M., Riesco, N., Jodar, I., and de Flores, T. (1992), "Denial mechanisms in myocardial infarction: their relations with psychological variables and short-term outcome," *Journal of Psychosomatic Research*, vol. 36, pp. 491-496.

Fields, K.B. (1989), "Myocardial infarction and denial," *Journal of Family Practice*, vol. 28, pp. 157-161.

Fishbein, M. (2000), "The role of theory in HIV prevention," Philadelphia: Public Policy Center, Annenberg School for Communication, University of Pennsylvania, unpublished work.

Fishbein, M., Bandura, A., Triandis, H.C., Kanfer, F.H., Becker, M.H., and Middlestadt, S.E. (1992), *Factors Influencing Behavior and Behavior Change: Final Report – Theorist's Workshop*, Rockville, Maryland: National Institute of Mental Health.

Foster, S. and Mallik, M. (1998), "A comparative study of differences in the referral behaviour patterns of men and women who have experienced cardiac-related chest pain," pp. 192-202.

Goff Jr., D.C., *et al.* (1998), "Knowledge of heart attack symptoms in a population survey in the United States," *Arch Intern Med,* vol. 158, pp. 2329-2338.

Gratch, J. (1999), "Why you should buy an emotional planner," in Velásquez, J. (Ed.), *Emotion Based Agent Architecture Workshop at Autonomous Agents'99 Conference*, www.ai.mit.edu/people/jvelas/ebaa99.

Hardin, P.C. and Reis, J. (1997), "Interactive multimedia software design: concepts, process, and evaluation," *Health Education & Behavior*, vol. 24, no. 1, pp. 35-53.

Horan,P.P., Tarborough, M.C., Besigel, G., and Carlson, D.R. (1990), "Computer-assisted self control of diabetes by adolescents," *Diabetes Education,* vol. 16, no. 3, pp. 205-211.

Isen, A. (1993), "Positive affect and decision making," in Lewis, M. and Haviland, J.M. (Eds.), *Handbook of Emotions*, Guilford Press, New York, Ch. 19.

Lajoie, S.P. and Derry, S.J. (1993), *Computers as Cognitive Tools*, Erlbaum, Hillsdale.

Lasseter, J., (1987), "Principles of traditional animation applied to computer animation," *Computer Graphics*, vol. 21, no. 4, pp. 35-44.

Lester, J.C., Converse, S.A., and Kahler, S.E., *et al.* (1997), "The persona effect: affective impact of animated pedagogical agents," *CHI 97*, vol. 22-27, pp. 359-366.

Lewis, D., (1999), "Computer-based approaches to patient education: a review of the literature," *JAMIA,* vol. 6, pp. 272-282.

Liao, L., Jollis, J.G., DeLong, E.R., Peterson, E.D., Morris, K.G., and Mark, D.B. (1996), "Impact of an interactive video on decision making of patients with ischemic heart disease," *Journal of General Internal Medicine,* vol. 11, no. 6, pp. 373-376.

Lister Hill Center (1998), *Symposium Proceedings, New Information Technology and the National Heart Attack Alert Program: Setting a 5-Year Agenda*, Lister Hill Center, National Library of Medicine, Bethesda, Maryland, April, pp. 14-15.

Magazine, M. (1999), "Remarks in his keynote address," 2nd *International Conf. On Automnomous Agents*, ACM Press, New York, (also see www.vpersons.com).

Marrero, D.G., Kronz, K.K., Golden, M.P., Wright, J.C., Orr, D.P., and Fineberg, N.S. (1989), "Clinical evaluation of computer-assisted self-monitoring of blood glucose system," *Diabetes Care,* vol. 12, no. 5, pp.345-350.

Meyerhoff, C., Bischof, F., and Pfeiffer, E.F. (1994), "Long-term experience with a computerized diabetes management and glucose monitoring system in insulin-dependent diabetic patients," *Diabetes Research in Clinical Practice,* vol. 24, no. 1, pp. 1-7.

Microsoft (1999), *Microsoft Agent Development Kit,* Redmond: Microsoft Press.

Norman, S. (2000), "Women's Contraceptive and Reproductive Experiences (CARE) Study."

Ortony, A., Clore, G.L., and Collins, A. (1988), *The Cognitive Structure of Emotions*, Cambridge: Cambridge University Press, Cambridge.

Osman, L.M., Abdalla, M.I., and Beattie, J.A., *et al.* (1994), "Reducing hospital admission through computer supported education for asthma patients: Grampian Asthma Study of Integrated Care (GRASSIC)," *British Medical Journal,* vol. 308, no. 6928, pp. 568-571.

Pennsylvania State Data Center (2000), *Pennsylvania Fact Sheet*, Penn. State Harrisburg Institute for State and Regional Affairs, 1-2.

Pincetl, P., Barnett, G.O., and Hoffer, E.P. (1987), "Chest pain: an exercise in clinical problem solving," *RxDx Software Series*, Williams and Wilkins, New York.

Psotka, J., Massey, L.D., and Mutter, S.A., (Eds.) (1998), *Intelligent Tutoring Systems – Lessons Learned*, Erlbaum, Hillsdale.

Reeves, B. and Nass, C. (1998), *The Media Equation: How People Treat Computers, Television, and New Media Like Real People and Places*, CSLI Publications, Stanford.

Reis, J. and Wrestler, F. (1994), "Consumer attitudes towards computer-assisted self-care of the common cold," *Patient Education and Counseling,* vol. 23, pp. 55-62.

Rhodes, F., Fishbein, M., and Reis, J. (1997), "Using behavioral theory in computer-based health promotion and appraisal," *Health Education & Behavior,* vol. 24, no. 1, pp. 20-34.

Rosenstock, I.M., Strecher, V.J., and Becker, M.J. (1994), "The health belief model and HIV risk behavior change", in: Diclemente, R.J. and Peterson, J.L., (Eds.), *Preventing AIDS: Theories and Methods of Behavioral Interventions*, Plenum Press, New York, pp. 5-24.

Rubin, D.H., Leventhal, J.M., and Sadock, R.T., *et al.* (1986), "Educational intervention by computer in childhood asthma: a randomized clinical trial testing the use of a new teaching intervention in childhood asthma," *Pediatrics,* vol. 77, no. 1, pp. 1-10.

Russo, J. (1998), "From pixels to personality: digital characters are virtually human," *Investors Business Daily.*

Silverman, B.G. (1998) "The role of web agents in medical knowledge management," *MD_Computing*, vol. 15, no. 4, pp. 221-31.

Silverman, B.G. (1995) "Computer supported collaborative learning (CSCL)," *Computers in Education Jnl.*, vol. 25, no.3, pp. 81-91.

Silverman, B.G., Cobbs, E., and Pincetl, P., *et al.* (1994), "An interactive learning environment for health care professionals,"

Proc. 18th SCAMC, American Medical Informatics Assoc., pp. 49-53.

Silverman, B., Cobbs, E., Pincetl, P., Motta, C., and Liao, R.L. (1992), "Toward an interactive learning environment for conveying biopsyhosocial-values skills to clinicians," GWU/IAI Tech Report.

Silverman, B.G. (1992a), "Human-computer collaboration," *Human-Computer Interaction*, vol. 7, no. 2.

Silverman, B.G. (1992b), *Critiquing Human Error: a Knowledge Based Human-Computer Collaborative Approach*, Academic Press, London.

Solloway, E. (Guest Editor) (1993), "Technology in education: special issue," *CACM*, vol. 36, no. 5.

Velásquez, J.D. (1999), "From affect programs to higher cognitive emotions: an emotion-based control approach," in Velásquez, J. (Ed.), *Emotion Based Agent Architecture Workshop at Autonomous Agents'99 Conference*, www.ai.mit.edu/people/jvelas/ebaa99.

Wener, E. (1997), *Artificial Intelligence and Tutoring Systems: Computational and Cognitive Approaches to the Communication of Knowledge*, Morgan Kaufman, Mountain View.

Wright, P., Milroy, R. and Lickorish, A. (1999), "Static and animated graphics in learning from interactive texts", *European Journal of Psychology of Education* (Special Issue on Visual Learning with New Technologies).

Chapter 10

Using Agents to Build a Practical Implementation of the INCA (Intelligent Community Alarm) System

M. Beer, W. Huang, and A. Sixsmith

This architecture is intended to integrate a number of autonomous systems; home monitoring, community alarms, care management systems and emergency systems command and control systems using agent technology to build effective coordinated care systems. A range of different autonomous bodies provides such care, many of which have their own management information systems already in place. Since these systems do not only contain information relevant to community care, but also all the other activities of the agent, that it would be unwilling to make available to other parties, the actual management of community care has remained primarily outside the role of current systems.

The development of an agent model is first described, followed by its realization using the ZEUS agent-building toolkit. While the toolkit approach allowed rapid development to take place, achieving a viable prototype in a very short time, various compromises had to be made with the original architecture. These are discussed, and the final demonstrator is described.

1 Introduction

Agent technologies have been proposed for a number of information applications (see, e.g., Jonker et al. 1999) in recent years with some success. The ability to build systems that allow information to be gathered effectively from various sources, each under its own autonomous control was shown effectively by the KRAFT project

(Preece et al. 1999). This provided a mechanism by which disparate heterogeneous databases and problem solvers could respond to each other through a common semantic network. Mediators that would route queries to the most appropriate information source handled query decomposition and recomposition. The information sources were connected to the network through wrappers that provided not only the physical connection, but also the semantic translation as well. This meant that there was a common semantic within the KRAFT network and that all messages could be routed anywhere. The project had a particular set of goals and it successfully showed that it could meet them.

Built on the KRAFT architecture Beer et al. (1999a) developed a similar approach for the management of interactions in a virtual college, and then for the development of a community alarm system (Beer et. al. 1999b). This chapter discusses the latter application, and the development of a prototype to demonstrate that a practical system could be developed using the principles developed. The practical implementation was built on the ZEUS agent building toolkit, which required considerable compromises with the original, theoretical architecture to meet the requirements of the toolkit methodology. Development has however been extremely rapid and a fully functional first prototype has been produced with about four months work.

Currently individual care providers use their own database, workflow and command and control systems with little or no integration between them. This causes difficulties, not only in the provision of the most effective response to emergencies, but also in the provision and management of routine care as the older person's requirements change, often quite rapidly. The current systems are not flexible enough to allow such changes to be implemented as rapidly as one would desire. The move towards franchising of different aspects of care delivery with the responsible agency (in the United Kingdom this is usually the Social Services Department of the Local Authority) being responsible for preparing a detailed specification, in the form of an Individual Care Plan. It then contracts various agencies to actually deliver the various components of the care, as appropriate. There is now no single agency with the overall authority to plan, manage, deliver and monitor the provision of community care. To this must be added the various health care services, and the emergency services, each of which have their own independent records and command and control structures, but are essential components in the delivery of the overall package. In addition, there are the large numbers of informal carers (family, friends, neighbors, etc.) who are currently almost totally ignored by the system, but provide invaluable support.

Proper use of the technologies available can give the necessary assurance to allow significant numbers of those currently requiring residential care to continue to live independent lives as part of their normal community. It is also necessary to show that this can be done without excessive external monitoring and data capture. The level of service must also be tailored effectively to meet the changing needs of all those concerned. This is currently a major issue, as increasing financial and resource pressures require much more active monitoring to ensure that care is directed effectively at those in most need.

2 Rationale

Recent years have seen a push to develop new technologies to support independent living and enhance patient care in the community. Developments include proactive, intelligent alarms that are able to raise alerts even when a person is incapacitated. Significant research and development effort has also been expended in fields such as clinical monitoring (Cellar et al. 1995), (Huang et al 1995), care planning and the application of video-telephony to enhance patient care (Sixsmith 1998). The imminent extension of effective high-speed network connections to a large number of UK homes gives many opportunities for the deployment of new technologies to support community care. These connections have considerable potential in:

1. replacing conventional technology, such as existing telephone-based services activated (as described by Fisk, 1989).
2. integrating specific functions into a single, comprehensive, framework for service delivery. The end users will then have a single access point to services, rather than the piecemeal approach currently adopted.
3. enhancing information sharing between all parties in a controlled and open manner. There are also many opportunities

to develop support for co-operative working.

4. developing new services, particularly those aimed at enhancing quality of life. These may include more general services from other service providers that also use the Internet as the delivery mechanism.
5. providing flexible, rather than dedicated, solutions, which are readily customizable to an individual's changing requirements.

The INCA system shows that current research into distributed knowledge sharing and intelligent agents (Wooldridge and Jennings, 1995) can be used effectively to provide a much higher level of support and care than is currently practical.

3 Related Work

Gerontechnology (Bouma and Graafmans, 1992) refers to the study of the interaction between older people and their technological environment. While this encompasses a wide range of design issues, the emphasis is on the need for the technical innovation to meet older peoples' needs, and in particular the use of state-of-the art technology to enhance their well-being. The areas of application encompass most aspects of life including, leisure, transportation, housing, social integration, health and work. Research and development in all these fields has the rationale of maintaining or improving the quality of life of older people by compensating for declining capacities and enhancing functional abilities.

Currently, community care is a labor intensive and low-tech activity characterized by face-to-face contact between carer and client. This situation is changing rapidly as products and services based on new technologies in areas such as information management, telecommunications and robotics become available. The difficulty has been that a large number of different agencies are involved with any individual, each with their own systems and procedures. Also, the agencies have to work with many different groups, including individual informal carers who do not have the sophisticated communications capabilities of the professionals. This means that many independent systems have to function together effectively in many different configurations. In a period of extremely rapid technological change, it is extremely difficult to conceive of a period when a common hardware, software and communications platform could be fully deployed by all involved in a cost-effective manner. Again, agent technology provides a potential method of integrating these different systems effectively, without excessive cost. Each agency can then communicate only the information required without having to pass on all the additional information that it may require for its own purposes.

Of course, community care is not entirely a technology-free zone. Most health and welfare professionals will be familiar with the alarm systems that enable older people to summon help by pushing a button on a telephone or on a pendant they wear around their neck. However, these kinds of alarms have significant limitations (Sixsmith 1994), particularly if the person is unconscious or is unable to push the button. Because of this, field trials of the 'Lifestyle Monitoring System' using sensors, such as infra-red movement detectors, are being carried out in a joint project between British Telecom, Anchor Housing and the Institute of Human Ageing. These are designed to see whether alarm systems can be enhanced to detect potential problems even when the person is incapacitated.

The difficulty is that there are a number of fiercely independent agencies involved in providing these services, each of which maintains its own information systems. These are specifically tailored to meet each organization's own individual needs, and any future system has to work effectively with all these existing 'legacy' systems.

4 Theoretical Approach

An agent mediated approach makes it possible to integrate these existing care systems and allows the INCA system to interface directly to a wide range of existing databases, knowledge bases and control systems that are already in place. The much broader care picture can thus be developed, and is of great help to care managers in developing the most effective and efficient care program for each individual, including the elimination of unnecessary duplication of resources.

Ensuring that the most appropriate assistance is provided in reasonable time will be a significant focus of the research. This will require multi-dimensional negotiation at a number of levels if acceptable levels of service are to be assured. The issue of timely intervention when delivery of the assistance promised fails will also have to be addressed, in that the appropriate agent must identify the anticipated failure at the earliest possible point. It must then attempt to renegotiate its commitment, either by finding another agent that will take it on, or by making a revised commitment. The process of negotiation is not as well founded in conversation theory as request and response exchanges, because there are much many more conditions that may apply. In the case of the INCA project these are simplified somewhat as it can be expected that all agents will use their best endeavors to meet all commitments that they enter into. This simplifies the analysis, without losing too much generality.

The provision of care inevitably requires a considerable degree of co-operative activity between individuals and agencies. Each actor is likely to have to interface with a number of different systems and organizational structures, each of which will have its own concept base (or Ontology) (Visser et al. 1999).

A major advantage of this approach is that the Home Unit (and sensors attached to it) can provide a range of services, rather than just one, as in the case of a Home Alarm or simple monitoring unit. Services can be added, modified and removed as required, providing a very high level of flexibility so that the changing requirements of the Older Person can be met in the most effective manner. This approach follows that outlined by Beer et al. (1999b).

4.1 The Problem Area Addressed

While community care covers a wide range of activities, five basic scenarios are considered in this chapter:

1. The development and updating of an Individual Care Plan. Care Planning is the specification of a "package" of care services according to an assessment of the needs of the individual.
2. The provision of routine care as specified by the Individual Care Plan and provided to an individual on a routine basis to support

day-to-day living.

3. The provision of positive care to maintain and enhance the quality of life.
4. Emergency support in response to some unexpected event, such as an accident or medical emergency.
5. The monitoring of the delivery of care to ensure that it is delivered as specified by the Individual Care Plan.

A number of organizations and individuals play a part in these activities. They can be grouped effectively into groups of actors, as shown in Table 1. The interactions between these actors can then be shown in the form of Use Case diagrams.

Table 1. The Actors.

NAME	DESCRIPTION
Care coordinator	The agency responsible for providing the range of services necessary to ensure that the Older person is properly cared for. The Care coordinator is responsible for preparing a Care plan and for monitoring its effectiveness in meeting the needs of the Older person. This is often the Local Authority or some other official body with a legal duty to provide the necessary care.
Care provider	The various agencies and individuals responsible for providing the care specified by the Individual Care Plan. This will include Social Workers, Health Care Professionals, Care Assistants, Emergency Services, Social Services etc. who can provide and extremely wide range of care services, if required.
Informal carer	The various relatives, friends, neighbors etc. who provide some form of support and assistance in an informal way (i.e. outside the Individual Care Plan), but is often essential to allow the older person to remain living at home. This is often flexible and responsive and can range from totally unstructured and so not recognized at all in the Individual Care Plan through to fully recognized and integrated with the efforts of the professional carers.
Older person	The person who lives in their own dwelling (either an ordinary house or a sheltered home) and who receives a package of community care services. This package may range from very minimal interventions, such as social alarm systems, through to an intensive mix of community support services.

4.1.1 Developing an Individual Care Plan

Before care can be provided effectively, the care coordinator has to assess need, and develop an individual care plan that meets the older person's requirements. This can be quite difficult, particularly with the severely disabled or those with dementia. If assessment is undertaken at an assessment center it ignores the specific circumstances and problems that occur in the home environment. However observation of a person in their own home requires considerable resources if an effective result is to be achieved. The monitoring facilities of the INCA system are a potentially invaluable source of information for such purposes.

It is also necessary to contact both formal care providers and informal carers to negotiate agreement to deliver the necessary care. The stages in doing this are shown in Figure 1.

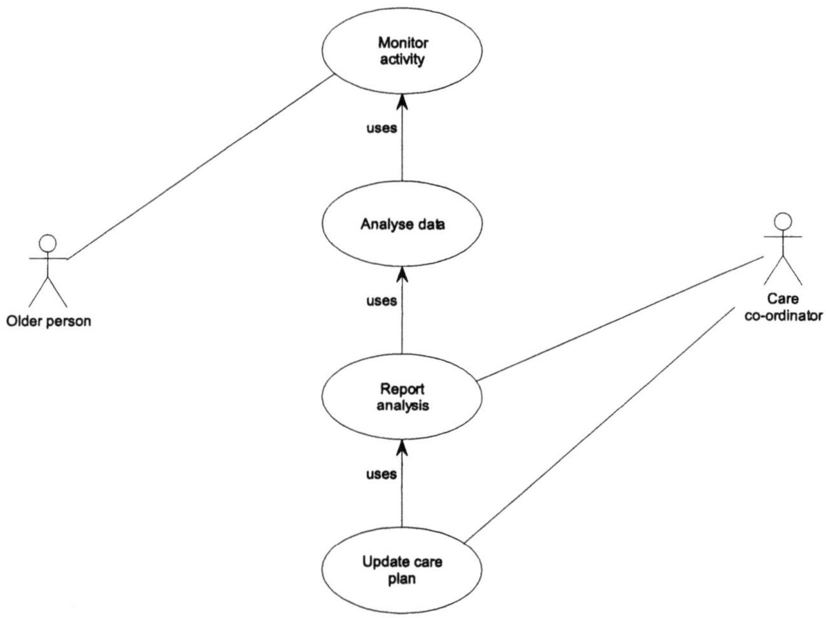

Figure 1. Individual Care Plan Development Use Case.

4.1.2 Positive Care

Once the Individual Care plan has been prepared, it has to be delivered. In general, Professional Care Providers actually deliver the care.

Informal carers are however extremely important in providing help and support, and need to be fully integrated into the care delivery system. Quality of life (or positive care) aims to improve the psychological and social well being of the Older Person. The INCA system facilitates this by:

1. Enhancing social interaction between the older person, informal carers and care providers
2. Providing information about activities and opportunities for leisure. These are actively filtered to give up-to-date and accurate information that is relevant to the Older Person.

Figure 2 shows the interactions involved in informing all concerned of the care required at any time.

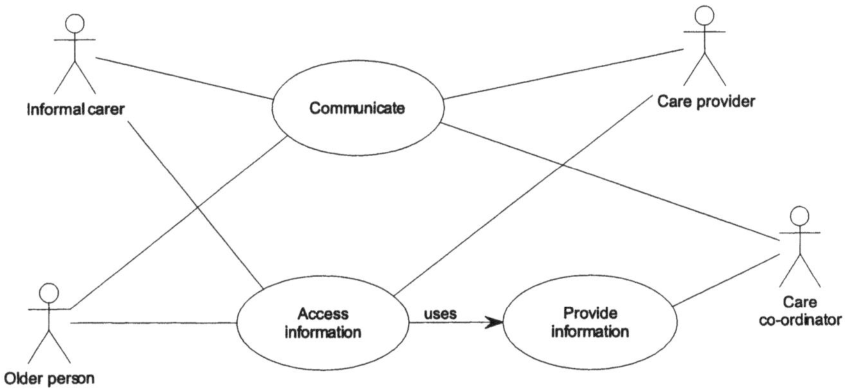

Figure 2. Positive Care Use Case.

The actual delivery of routine care has to be fully monitored against the Individual Care Plan. The objective is to provide the assistance necessary to make up for the disabilities of the individual, such as help with eating, washing, assistance with bodily functions and getting in and out of bed. The main problem with existing systems is the lack of responsiveness of the services providing it. In high quality residential settings this kind of care is delivered whenever the person requires it. In a community setting, this care is usually provided on a very rigid schedule unless informal carers are willing to take on a very large burden. The INCA system is designed to enhance the responsiveness of the service by:

1. Allowing users to request care directly from care providers and coordinators.
2. Providing users with better information about care schedules, such as when a care provider will arrive, delays and changes to schedules at the earliest possible opportunity.
3. Monitoring very disabled people (such as people with dementia) in order to fit care interventions more closely to their patterns of daily activity.

Figure 3 shows the interactions involved.

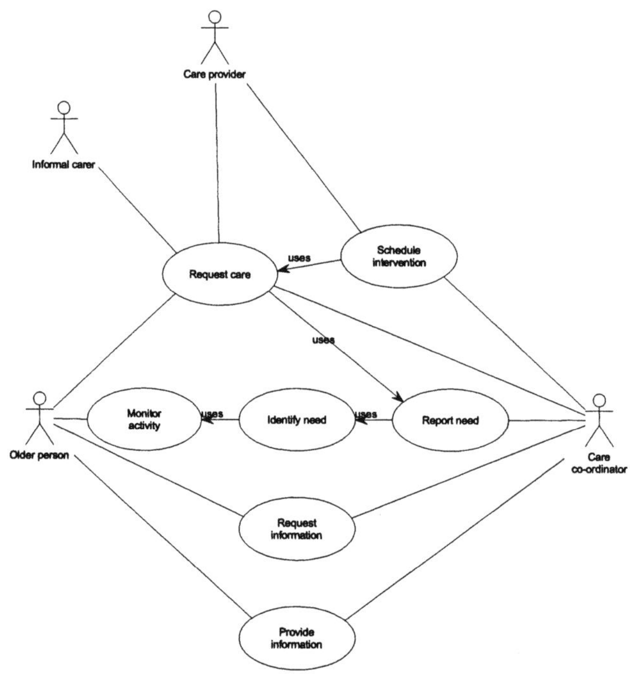

Figure 3. Routine Care Use Case.

4.1.3 Emergency Support

The case of emergency intervention is rather different. The intention is to provide a proactive alarm system, in addition to the current reactive systems that operate independently of the overall care provision. The primary interface is through the home unit that raises an alert either when some set of environmental conditions is considered out of range, as defined within the individual care plan, or there has been some direct request for help (such as pressing a 'panic button'). These interactions are shown in Figure 4.

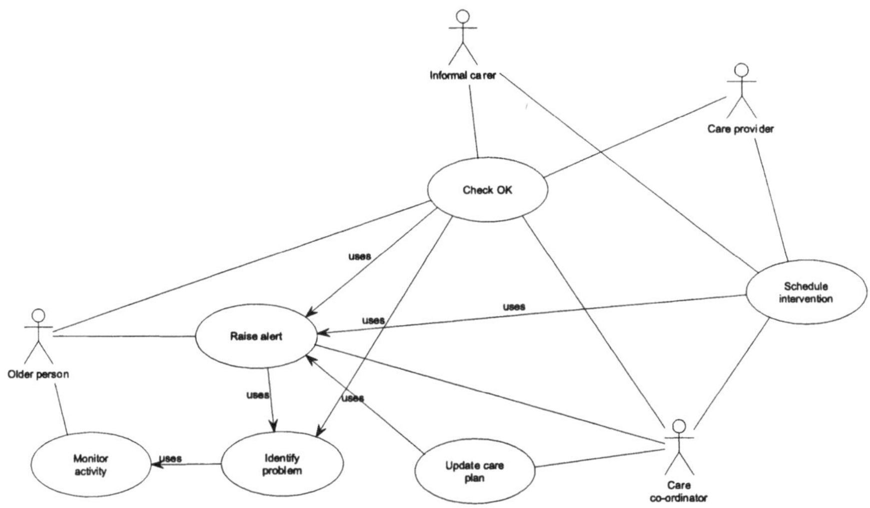

Figure 4. Emergency Support Use Case.

4.2 Care Management

Once the Individual Care Plan has been prepared, the requirements of the Older Person have to be continually monitored so that the plan can be updated as necessary so that the requisite level of care continues to be provided. The Care Coordinator, as shown in Figure 5, undertakes this activity. Key to the INCA concept is that the care plan should be dynamically updated to meet current needs, rather than simply designed to meet those identified at the original assessment. The ability of the system to continuously monitor both requirements and service delivery is key to this.

It is also important to ensure that all the care specified in the Individual Care Plan is actually delivered to acceptable standards, at the appropriate time and in the right place. Monitoring is problematic in the community context, as direct supervision of care staff is almost impossible. The INCA system facilitates effective monitoring in two ways:

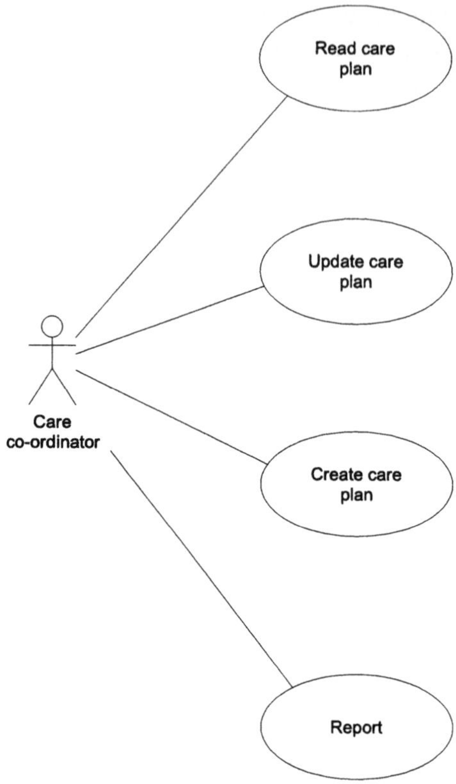

Figure 5. Preparing and updating the Individual Care Plan.

1. Care providers log their interventions directly into the system at each visit. These can then be compared directly with the contents of the Individual Care Package. Any deviations can then be investigated and either the Individual Care Package can be updated or other appropriate action undertaken.
2. Complaints procedures can be based on direct communication to the Care Coordinator, improving monitoring and responsiveness.

Figure 6 shows the interaction involved in Quality Assurance procedures.

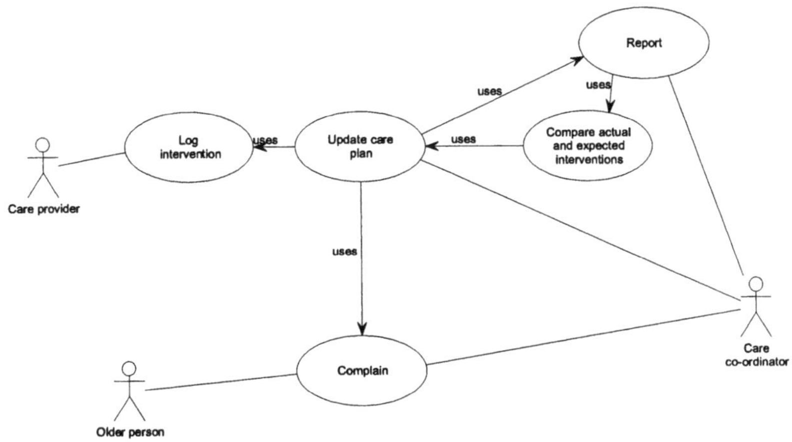

Figure 6. Quality Assurance Use Case.

Once the facilities required for supporting community care have been provided, they can be used with little or no additional hardware to provide additional useful services. For example, if it were programmed to act as a burglar alarm when the house is empty, it would ask the facilitator for a different service mediator which handles burglar alerts, and a different set of responses would be triggered, that would be appropriate to that situation. This would be implemented by activating a different set of conversation classes, with their own guard conditions, post conditions and completion conditions. In this way, the operation of the 'burglar alarm service' can be kept separate and independent from the 'community alarm services'. Only the requirements of the various community alarm services are considered here, as these need to be kept independent from each other, and additional services are simply extensions of the architecture proposed.

5 The Design of the Conversation Classes

The previous section shows that a wide range of activities can be supported effectively by the INCA system. Traditional database design techniques have difficulties with this richness, as each scenario contains actors with the same roles performing differently depending on their beliefs and intentions within that scenario. Agent approaches are specifically intended to address these issues. This can be illustrated effectively by considering the management of communication for dealing with emergencies.

An alarm condition is raised only when the home unit detects sensor readings outside the normal range. A facilitator is contacted to obtain the address of a suitable service mediator, which is then alerted. This mediator routes the alert to a suitable service provider after adding additional information from the service database. The service provider's mediator determines the appropriate course of action and notifies the necessary carer, possibly via further mediators. This allows considerable additional flexibility, as for example, the Warden may only be available during certain hours, on certain days of the week. A mediator can ensure that all alerts are appropriately routed to obtain an immediate response.

One of the main features of the work described is the specification of the conversational abilities of agents. In line with KQML (Lebrou, 1996), the conversational abilities of an agent will be determined by a set of speech acts, which that agent can perform, and a set of speech acts to which it can respond. *Speech acts* are used here in the familiar way, deriving from Searle (Searle, 1969). A speech act comprises a *performative*, indicating the illocutionary and intended perlocutionary force of the act, together with an *argument*, typically a proposition.

The semantics for the speech acts are provided as follows, following Lebrou (Lebrou, 1996):

1. a set of pre-conditions for the performance of the act,
2. a set of post-conditions to be enforced immediately on performance of the act, and
3. a set of completion conditions which are to apply when all the intentions associated with the act have been finally satisfied.

This means that the completion conditions may be achieved at some time after the communication being described has taken place, after some further conversation has occurred. This means that conditions or acts with such conditions can provide a context for later utterances. Performatives provide the building blocks for conversations between agents. However, for conversations to take place, they must be coordinated in a way appropriate to the particular conversation.

Barbuceanu et al. (Barbuceanu et al. 1995) show the necessity for this level of coordination. In that paper, conversations are specified by using state transition diagrams, but greater uniformity can be achieved by extending Labrou's method and representing the conversation rules as additional preconditions, post-conditions and completion conditions on the speech acts they use. Experimental use of this approach has been reported in (Bench-Capon, 1998), which describes the specification of conversation rules for several dialogue games. Thus far the performatives are specified in a way that is neutral as between the agents that perform and respond to them. In our particular application, however, there is a need to customize the act to particular agents. It is essential for confidentiality reasons that an agent should be required to respond to questions from some particular class of agents only under particular conditions. Such constraints can be included uniformly in the specification above by including pre and post conditions local to individual agents expressing heuristics and policies. For example:

1. Personal information must only be provided to those involved in the provision of care to that individual and then only when it is required in the provision of that care.
2. Communication between agents that is only relevant to those agents must be handled appropriately. Examples of this would be messages between the care coordinator and the elderly person to check the level of care being provided must know who is responding. Another example would be where the care coordinator is scheduling the carer. This will usually involve information about a number of elderly people that the carer is scheduled to visit.

The basic mechanism for the home unit to communicate with service providers is shown as a transition state diagram in Figure 7. This shows the actions required summoning assistance. The actual service provider summoned depends on a number of factors:

1. The type and severity of the problem
2. The service providers able and willing to respond
3. The anticipated speed of that response
4. The 'cost' of that response in relation to the perceived need

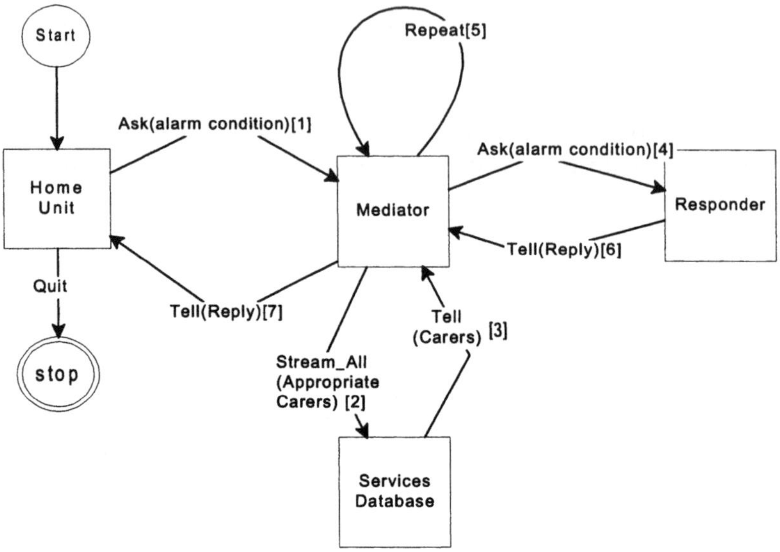

Figure 7. The Basic Response Conversation Class.

The first is initially controlled by the guard conditions on the conversation class. (Gray et al, 1997) and is determined by information local to the Home Unit (the reason for the alarm, whether the home is occupied or unoccupied etc.). Decisions based on the service provider's ability or willingness to respond require much more general knowledge that is available to the mediator, which can negotiate with a number of service providers to determine whether one or more can respond in an appropriate way within a reasonable time. If a service provider is able to do so, then the mediator accepts the most appropriate offer, rejecting all others, and informs the Home Unit. In certain cases no single service provider would be appropriate. If, for example, a serious medical emergency was detected and an ambulance would arrive only after some time, the mediator would ensure that a carer closer to the scene was also called and would look after the older person until help arrived. This would require several sets of negotiations and many interactions before a satisfactory result was achieved.

The much broader care picture can thus be developed, and is of great help to care managers in developing the most effective and efficient care program for each individual, including the elimination of unnecessary duplication of resources.

A major advantage of this approach is that the Home Unit (and sensors attached to it) can provide a range of services, rather than just one, as in the case of a Home Alarm or simple monitoring unit. Services can be added, modified and removed as required, providing a very high level of flexibility so that the changing requirements of the Older Person can be met in the most effective manner.

6 The Benefits of Using Agents for INCA

The integration of existing care systems, facilitates the integration of a wide range of existing databases, knowledge based and control system programs. Agent technologies enable a 'controlled' degree of co-operative activity between these independent components. They are 'Controlled' in the sense that each component within the system be it an organization or information store need only communicate the information required by the system. There should be no requirement to open additional information required for own proposes. Therefore, taking an agent-based approach it is possible for the organizations involved in the provision of care to communicate effectively yet securely, plus have access to much needed information. All of which will lead to a more managed and effective approach to care provision thus providing an increased speed of service and the elimination of unnecessary duplication of resources.

The ability of agent architectures to be distributed across heterogeneous environments has enabled the development of the INCA project's care system integration methodology. In the past a barrier to coordinated community care has been the incompatibility of the independent agencies systems and procedures. Agent-based methodologies like INCA enable heterogeneous systems to function together effectively in many different configurations. Thus agent technologies provide the means to integrate the different health care agency systems regardless of organizational structure and provide the means for the various agencies to interface on a common domain level.

Within the scope of the term 'agent technologies' falls a wide range of different agent typologies, for example agents may be information agents, reactive agents or collaborative agents. In the context of the

INCA architecture, the agents within the system are collaborative agents.

The collaborative nature of an agent places an emphasis on autonomy and cooperation with other agents.

> *Autonomy* means that agents have the ability to act without the need for human invention and guidance. They are able to achieve goals on behalf of the user and the system in a pro-active manner rather than a re-active manner. In other words, agents are able to take the initiative.
> *Co-operation* means that agents are able to co-operate with other agents within the architecture in order to achieve domain goals.

The rationale for constructing an architecture of collaborative agents is that numerous collaborative agent provide solutions to inherently distributed problems and are suited to solving problems that are too large for a single centralized agent. Collaborative agents negotiate with the other agents within the system thus organizing and accomplishing complex goals and tasks that are often beyond the capabilities of a single agent.

It will be seen that the INCA architecture enhances the co-ordination between the various health care agencies by enabling the agents within the system-controlled access to the information sources incorporated within the INCA domain. That is, it is essential that the agents within the INCA architecture can interface with the existing agency information sources, including sources such as databases, log files and knowledge bases.

In order to facilitate effective co-ordination, it is necessary to allow the agent within the architecture to read and write domain information to and from the appropriate organizational information source. This can be achieved by the use of data wrappers, which provide an interface between agent and information source. Data wrappers enable the agents to query (read) information, by translating agent requests constructed in the communication protocol of the INCA domain into a form that is understood by the agency's own information system. For example an agent request may be translated into a SQL statement that can be executed against an organizational database. In the same manner, the responses (provided by said information sources) to agent requests are translated or mapped back into the agent communication protocol of the INCA domain.

7 Constructing the Demonstrator

7.1 Developing Conversation Class Model Using UML

Agent software engineering is currently very immature, and a number of frameworks have been defined (see Ciancarini & Wooldridge, 2000). A number of different methodologies have been proposed, but none are fully satisfactory for developing complex agent-based applications. In this case an approach based on UML (Baur et al. 2000) was adopted as this fitted well with out earlier approach in developing the conversation class model. Following this approach, a collaboration diagram for the INCA system was produced, as shown in Figure 8.

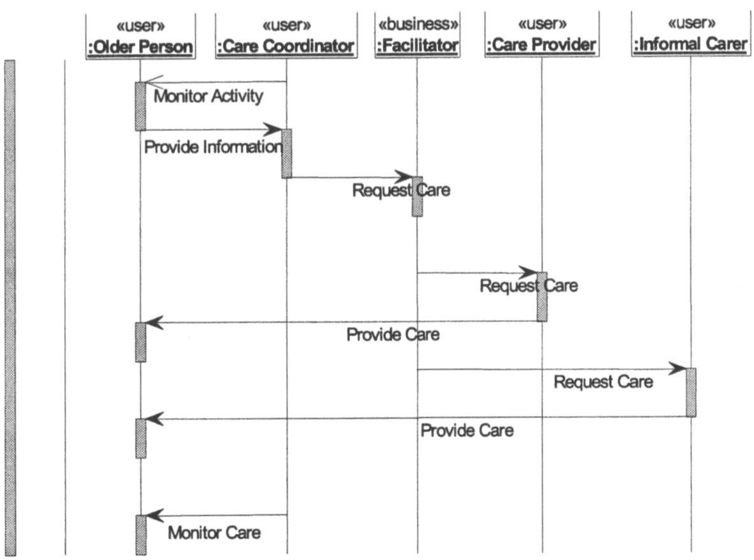

Figure 8. Collaboration diagram for the INCA System.

The agent-based demonstrator implements an integrated health care system and therefore simulates the utilization of various health care resources. Resources that are produced acquired and consumed by actors/agencies within the INCA domain. The provision of health care involves the acquisition (by a care coordinator) of a number of different care resources and services, which are supplied by a number of different independent organizations or actors. Ultimately, the actors that represent the older people within the community care system, consume these acquired resources. Whilst considering the above description of the problem domain, work began on the agent architecture's design stage.

7.2 Phases of the ZEUS Design Methodology

The detailed design for the INCA prototype was structured by the design methodology expounded by the developers of the ZEUS toolkit (Nwana et al. 1999). The phases of the ZEUS design methodology are as follows:

Phase ONE – Agent Identification

Best agent candidates are the following entities:

AUTONOMOUS: can function without humans.
SOCIAL: capable of interacting intelligently.
RESPONSIVE: can respond to their environment.
PROACTIVE: capable of taking initiative.

E.A Kendall (1997) states that the actors defined within a use case diagram depicting an agent architecture may be naturally replaced by agents that augment the collaboration between them. In other words, software agents map onto actors. A number of generic agents were therefore identified for inclusion within the INCA architecture, based on the use cases outlined on Section 4.

The following task agents are to be found within the INCA demonstrator.

CARE COORDINATOR: represents the agent responsible for the overall provision of care. The care coordinator formulates the Individual care plan on which care provision is based and is also charged with monitoring the effectiveness of the care provided.

CARE PROVIDER: represents the generic form of the various actors/agencies that supplies the health care resources and services as specified within an older person's Individual care plan.
INFORMAL CARER: represents the friends and family of the older persons within the health care system, thus giving the informal carers a formal means of representation within the INCA domain. The Informal carer can supply health care resources under the flexible term 'Support'.
OLDER PERSON: represents the older person who receives and consumes the health care services and resources.

The provision of care is within the 'real-world' environment, is normally supplied by *a number* of different independent 'Care Providers'. To design the INCA demonstrator with just one care provider seemed a little limiting. Therefore three additional 'Care provider' agents were also envisaged. The agent identified as the 'CARE PROVIDER' was taken to characterize a generic care provider, and could represent any of the health care organizations responsible for supplying care. The additional agents are as follows:

EMERGENCY SERVICES: represents the ambulance, police and fire services and is responsible for the provision of care as required during an emergency situation.
DOCTOR: represents a General Practitioner or more generally any generic 'Doctor' envisaged role.
NURSE: This agent represents a distinct nurse or any other generic nurse role.

Phase TWO – ZEUS Role Model Selection

The use of role models allows the developer to structure the overall development of an agent architecture by applying the application specific problem (i.e. the INCA domain) to a pre-defined model of an agent system. In this way, the developer can visualize the system within the context of the selected role model and organize agent design in terms of role fulfillment and role based responsibilities as defined within the selected role model.

After careful consideration the role model chosen was the 'Simple Supply Chain'. The simple supply chain comprises of a number of suppliers (who can supply resources on demand) and consumers (who consume the resources once acquired from the suppliers). This role model appeared to map on to the INCA domain with relative ease. The INCA domain has a number of suppliers in the form of the health agencies that supply the care services and resources and a number of consumers, these being the older persons who receive emergency services and the care coordinator who obtains resources on behalf of the older persons.

Utilizing the simple supply chain enables an agent to be classified as a supplier or a consumer. Associated with this classification, as supplier or consumer, are a number of 'roles and responsibilities' that an agent needs to fulfill. For example if an agent is classified as a consumer, the agent needs to fulfill the roles of 'Negotiation Initiator' (thereby requesting a resource to be supplied) and 'Consumer' (consuming the resource once obtained). On the other hand if an agent is a supplier, it needs to fulfill the roles of 'Negotiation Partner' (negotiating whether to agree or refuse a consumer request for a resources), 'Producer' (producing the required resource if needed) and 'Supplier' (actually sending the resource out the consumer). The ZEUS toolkit automatically provides the social\agent level functionality required in order for the agents within the architecture to fulfill their associated roles and responsibilities as suppliers and consumers, thus allowing the developer to concentrate on domain specific task related issues. This is shown diagrammatically in Figure 9.

Phase THREE – Ontology Creation

The next phase in the agent architecture design process is the creation of a domain vocabulary or ontology that represents the concepts of the INCA domain that are shared by all the agents. For example in a fruit trading domain the concepts involved could include physical items such as apples, oranges and pear and more abstract concepts such as money. Within the INCA architecture, the concepts that form the domain ontology are the health care resources produced and consumed by the agents within the system.

In order to communicate agents require both a common language and a common shared representation of domain concepts. A common domain ontology must therefore be created. Within the ZEUS toolkit methodology terms included in an ontology are known as ZEUS facts. ZEUS facts represent the resources that are utilized within an agent system. The concepts that form the main part of the domain ontology are the health care resources produced and consumed by the agents within the system. For example, in the real world of health care provision, the emergency services can provide fire, ambulance and police resources. Therefore if these resources are to be utilized within the INCA domain, they must be represented as ZEUS facts. That means that ZEUS facts called 'Ambulance', 'Fire' and 'Police' will need to be created.

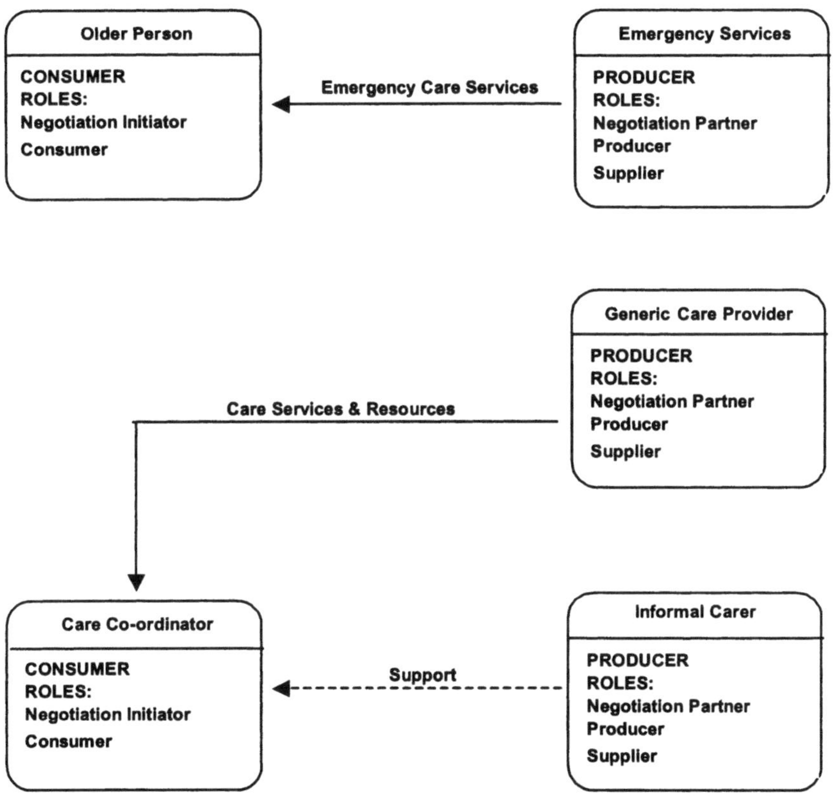

Figure 9. The INCA Supply Chain.

The INCA ontology was continually changed during the implementation stage of the project as new more abstract ZEUS facts were needed in order to implement agent task functionality. For example, if an agent required information from a database, it was necessary to create a ZEUS fact to represent the information read from the database, so that it was available to the agents within the system. This mechanism is shown in Figure 10.

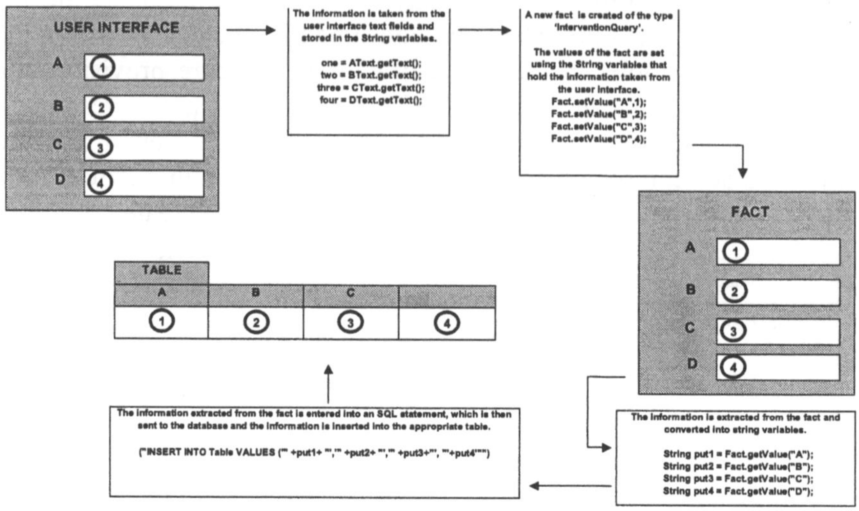

Figure 10. Integration data input through Interface Agents.

Phase FOUR – Agent Definition

The following questions must now be answered:

- What tasks can the agent perform?
- What initial resources are available to the agent?

First, the agent's main *domain* level task has to be defined, for example that it is required to acquire the health care services and resources needed by the older persons as specified in an Individual Care Plan. This task needs to be given a task name, in this case 'Request_Care'. At this stage of agent design, the developer is not required to specify the task in detail, it is enough to state that this agent can perform this task). This agent does not possess any initial resources, as it obtains any resources required from the *supplier agents* within the INCA system by means of it's 'Request_Care' task.

Phase FIVE – Task Definition

During this stage of design the details of agent domain tasks are specified with regard to the task's preconditions, its effects, how much it costs, how long it takes and any constraints. The task 'Request_Care' in the context of the demonstrator has no cost, duration or constraints, however it does have a number of preconditions (Inputs) and an effect (Output).

The task 'Request_Care' takes as its' preconditions the health care services and resources (represented by ZEUS facts) needed in order to produce an Individual Care Plan (again represented by a ZEUS fact). Once these Facts have been obtained from the appropriate care provider agents an Individual Care Plan ZEUS fact could be specified.

Phase SIX – Agent Organization

During this stage the developer must identify for each agent, other agents that are known within the system and identify the relationship between each known agent. Next the developer needs to define what each agent knows about other agent's abilities. Figure 11 shows the overall architecture with these interrelationships.

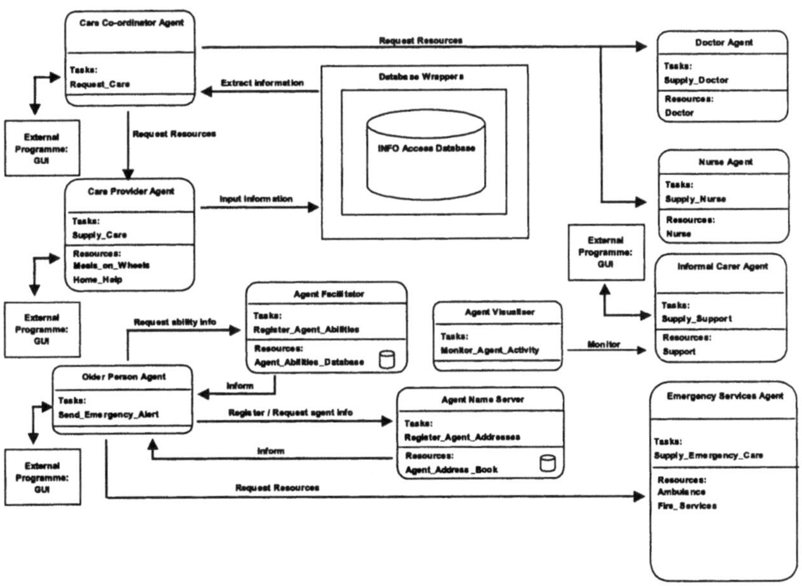

Figure 11. The Overall Architecture of the INCA Prototype.

Phase SEVEN – Agent Coordination

During this phase the developer needs to choose from the ZEUS library of co-ordination protocols. These protocols define how agents interact and negotiate with other agents within the architecture. The ZEUS design methodology is organized so that once all the details of the agent architecture have been obtained, the details of the architecture can be entered into a number of editors provided by the ZEUS toolkit and Java source code can then be automatically generated. This was undertaken successfully and the overall architecture shown in Figure 12 was the result.

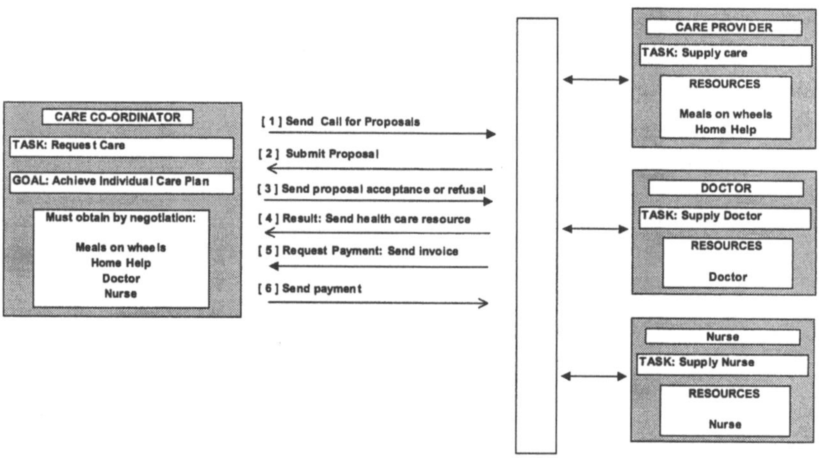

Figure 12. The negotiation mechanisms adopted by the Care Coordination agent to obtain care resources.

7.3 User Interface Design

The design for the graphical user interfaces was based on the interface design as expounded by the ZEUS toolkit's developers. The components considered as necessary design features were as follows:

- An area in which to view inter-agent communication and system wide messages. For example, this area could display the agent conversations that take place during agent negotiation.
- Buttons that trigger agent tasks/goals, such as to request care resources in order to construct an Individual Care Plan, and

initiate agent requests for services and information.

- Text areas in which to enter and view information, such as to compose messages or to enter or retrieve information.

Figure 13 gives a typical simple interface, that of the home unit. The main area displays messages based on information passed to and from the other agents. The buttons along the bottom of the window provide the controls. The 'Send Alert' button initiates a request for assistance, based on the care plan and the agent's current knowledge. There is also the capability to send messages to the Care Coordinator agent by pressing the 'Enter Message' button. A supplementary text window then appears below the main window, in which the message can be entered. It is sent when the 'Send Message' button is pressed.

Figure 13. The User Interface for the Older Person.

8 Implementation

During implementation the benefits of utilizing the ZEUS toolkit in developing the INCA demonstrator become all too clear. Once the INCA architecture had been defined, using the ZEUS design methodology, the information regarding agent configuration could be entered into a number of graphical editing windows supplied as part of the ZEUS toolkits 'Agent Generator' tool.

8.1 The Agent Realization Process

The agent realization process closely followed the stages of the ZEUS agent design process, the following is a brief overview of the implementation of agent/social functionality (which was achieved by utilizing the ZEUS Agent Generator tool).

Ontology Creation: The application Domain ontology represented as ZEUS 'Facts' is entered into the ZEUS Ontology Editor.
Task Agent Creation: The Information relating to design considerations is entered into agent construction editors.
Task Configuration: The domain level tasks (such as 'Request_ Care') are specified within Task Editors.
Utility Agent Configuration: The architecture's Utility agents (that provide architectural infrastructure support) are specified within the ZEUS Code Generation Editor
Task Agent Configuration: The architecture's task agents are specified, as to the external data sources they are connected to (that is the INCA demonstrator's Access database) and the external programs they are connected to (in this case the GUI programs implemented in Java Swing). This information is also entered in the ZEUS Code Generation Editor.
Agent Implementation (Code Generation): During this final stage, the ZEUS Code Generator automatically generates Java source code. This generated Java code implements the social level agent functionality needed in order for agents to co-operate and interaction within a collaborative architecture.

Once the social/agent level code has been generated and implemented, the next stage of the realization process is for the developer to implement code for task activation, the external GUI programs and external resource connectivity.

8.2 Implementing the Graphical User Interfaces

The Graphical user interfaces required for the INCA demonstrator were implemented in Java Swing (an extension of the Abstract Windowing Toolkit available with Java 1.2). The GUIs connected to the INCA architecture (via a Java interface) enabling the user to send and receive information to and from the agent system.

The user interfaces main functions were as follows:

1. To instruct the agents to achieve a certain goal or task. For example, if the 'Request Care' button is pressed on the User Interface connected to the Care Coordinator agent, the agent would attempt to obtain the necessary care resources in order to produce an Individual Care Plan.
2. To allow the user the means of entering information into a 'data form' for storage within the Access Database integrated with the INCA agent architecture.
3. To allow the user to monitor and view the conversations generated due to agent negotiation.

9 Further Work

The work described here provides a solid foundation for future work regarding the INCA project. The practical side to the INCA project can be developed to further incorporate the informal health care providers within the INCA domain (such as friends, family and neighbors) and to provide better communication channels between the actors, agencies and organization integrated within the INCA domain. There are significant opportunities to implement such services using for example WAP mobile telephones. Such communication devices provide many opportunities to improve communications capabilities to both care workers in the field, and to those who are currently left out of the formal structures, such as the informal careers.

The INCA demonstrator successfully showed that the INCA agent architecture is a feasible proposition at least on a laboratory scale. It is now necessary to develop a fully operational pilot to show that the technology is able to cope fully with the vagarities of practical networks. It is also necessary to show that the wrapper-based technology can be used effectively to provide properly controlled interfaces to real management information systems. Only when this has been shown to work under full operational conditions will care managers and other key policy makers have full confidence in this technology. This is essential for successful adoption both in terms of its effectiveness and the security of what is often very sensitive data. The later can be addressed effectively by using the encryption techniques currently being developed for electronic commerce applications. It is clear that as agent technologies continue to the possibilities for agent-based applications seem only to be limited by the ever-increasing knowledge base and imagination of the developer.

The current state of agent communications technology has required that significant compromises have been made in the communications infrastructure. In particular, the negotiation of prices, and mechanism for authorizing payment for emergency care is quite primitive, being based on a simple auction model. This is an area that urgently needs more analysis, so that proper business-based protocols can be developed. The issue here is that the party requesting and authorizing the necessary care is not that which will eventually pay for it. This is common in business transactions but has not been fully developed in the agent literature yet.

It could be suggested that the informal carers within the INCA domain were not fully integrated within the demonstrator, as they were not implemented within the supply chain (for supplying health care resources required in order to produce an 'Individual Care Plan'). The demonstrator did however provide a means to formally represent the informal carers within the INCA domain and also provided the means to incorporate them with a supply chain scenario. Informal carers can

and do supply the health care resources loosely known as 'Support', it would require little modification to the supply chain represented by the demonstrator, to fully incorporate the informal carer. As the informal carer supplies health care resources (and as such represents a generic care provider) the 'Informal Carer' agent could replace any of the 'Care Provider' agents within the current supply chain.

10 Conclusions

The INCA demonstrator shows how agent-based applications can be developed from initial analysis through to a working laboratory demonstrator. A major result has been the implementation of the agent negotiation component of the INCA demonstrator, in which the collaborative agents within the architecture co-operated in order to accomplish the goal of the provision of health care services/resources (whether routine or emergency). This demonstrated that the utilization of agent-based technologies could be undertaken in order to achieve distributed demand and supply issues within a integrated domain space, whilst keeping the actors and agencies (within the domain) independent, with full recognition of each actor's jealously guarded autonomy. The implementation of the agent negotiation component as modeled on the 'Simple Supply Chain' model was deemed a greater success as the example application (based on the 'Simple Supply Chain') included with the ZEUS package failed to operate correctly.

The successful implementation of the data wrapper approach to the integration of an external data source, with the INCA agent architecture provides valuable support for the INCA projects integration program. The demonstrator showed that it is possible to integrate data sources as part of an overall agent system, data sources such as those associated with the various health care organizations incorporated within the INCA domain.

11 Acknowledgments

Iain Anderson as part of his MSc project developed the initial prototype program for the agents in the summer of 2000. We are also grateful Trevor Bench-Capon for his comments and assistance with speech acts.

12 References

Bauer, B., Muller, J.P., and Odell, J. (2000), "Agent UML: a formalism for specifying multiagent software systems," in Ciancarini, P. and Wooldridge, M.J., *Agent-Oriented Software Engineering*, Lecture Notes in Computer Science 1957, Springer.

Bouma, H. and Graafmans, J. (1992), *Gerontechnology*, Amsterdam. IOS Press.

Beer, M.D., Bench-Capon, T.M., and Sixsmith, A., (1999a), "Dialogue management in a virtual college", *Proceedings of Dexa'99*, Florence, Italy, August, (Lecture Notes in Computer Science 1677), Springer-Verlag, pp. 521-530.

Beer, M.D., Bench-Capon, T., and Sixsmith, A. (1999b), "The delivery of effective integrated community care with the aid of agents", *Proceedings of ICSC'99*, Hong Kong, December, (Lecture Notes in Computer Science 1749), Springer-Verlag, pp. 303-398.

Bench-Capon, T.J.M. (1998), "Specification of communication between information sources', *Proceedings of DEXA98*, Lecture Notes in Computer Science, Springer, Berlin.

Celler, B.G.,Earnshaw, E.W., Ilsar, E.D., Betbeder-Matibet, L., Harris, M.F., Clark, E.D., Hesketh, T., and Lovell, N.H. (1995), "Remote monitoring of health status of the elderly at home: a multi-disciplinary project on ageing at the University of NSW," *Int. Journal of Biomedical Computing,* vol. 40, pp. 147-155.

Ciancarini, P. and Wooldridge, M.J. (2000), "Agent-oriented software engineering", *Lecture Notes in Computer Science 1957*, Springer, Berlin.

Finnin, T., Labrou, Y., and Mayfield, J. (1997), *KQML as an Agent Communication Language,* MIT Press, Cambridge MA.

FIPA (1997), "FIPA 97 Specification," Foundation for Intelligent Physical Agents, available from http://www.fipa.org/.

Fisk, M. (1989), *Alarm Systems and Elderly People*, Planning Exchange, London.

Gray, P.M.D., Preece, A., Fiddian, N.J., Gray, W.A., Bench-Capon, T.J.M., Shave, M.J.R., Azarmi, N., Wiegand, M., Ashwell, M., Beer, M.D., Cui, Z., Diaz, B., Embury, S.M., Hui, K., Jones, A.C., Jones, D.M., Kemp, G.J.L., Lawson, E.W., Lunn, K., Marti, P., Shao, J., and Visser, P.R.S., (1997), "KRAFT: knowledge fusion from distributed databases and knowledge bases," *Database and Expert System Applications (DEXA'97)*, Toulouse, France.

Huang, J., Jennings, J.R., and Fox, J. (1995), "An agent-based approach for distributed care management," *Applied Artificial Intelligence: an International Journal*, vol. 9, no. 4, pp. 401-420.

Jones, D.M., Bench-Capon, T.J.M., Visser, P.R.S., Diaz, B., Beer, M.D., and Shave, M.J.R. (1999), "Resolving ontological heterogeneity in the KRAFT project", *Proceedings of Dexa'99*, Florence, Italy, August (Lecture Notes in Computer Science 1677), Springer-Verlag, pp. 668-677.

Jonker, C.M., Lam, R.A., and Treur, J. (1999), "A multi-agent architecture for an intelligent website for insurance", *Proceedings of Cooperative Agents III*, (Lecture Notes in Artificial Intelligence 1652), Springer, pp. 86-100.

Kendall, E.A., Malkoun, M.T., and Jiang, C.H. (1997), "The application of object-oriented analysis to agent based systems," The Report on Object Oriented Analysis and Design in conjunction with *The Journal of Object Oriented Programming*, February.

Labrou, Y. (1996), "Semantics for an Agent Communication Language," Ph.D. Thesis, University of Maryland, Baltimore.

Nwana, H., Ndumu, D., Lee, L., and Collis, J., (1999) "ZEUS: a tool-kit for building distributed multi-agent systems," *Applied Artificial Intelligence Journal*, vol. 13, no. 1, pp. 129-186.

Preece A.D., Hui, K.-Y., Gray, W.A., Marti, P., Bench-Capon, T.J.M., Jones, D.M., and Cu, Z., (1999) "The KRAFT architecture for knowledge fusion and transformation," *19th SGES International Conference on Knowledge-Based Systems and Applied Artificial Intelligence (ES'99)*, Springer, Berlin.

Sahuguet, A. and Azavant, F. (1998), "W4F: the WYSIWYG Web wrapper factory," Technical report, University of Pennsylvania, Department of Computer and Information Science, http://cheops .cis.upenn.edu/sahuguet/WAPI/.

Searle, J.R (1969), *Speech Acts: an Essay in the Philosophy of Language*, Cambridge University Press.

Shave, M.J.R. (1997). "Ontological structures for knowledge sharing," *New Review of Information Networking*, vol. 3, pp. 125-133.

Shin, A.P.J. (1998), "INAS – Intelligent Alarm System for the Elderly," Honours Project Dissertation, University of Liverpool.

Sixsmith, A. (1994), "New technology and community care," *Health and Social Care*, vol. 2, no. 6, pp. 367-370.

Sixsmith, A.J. (1998),"Telecare at home," in Walker, A. (ed.), *European Home and Community Care 1998/99*, Campden, London.

Wiederhold, G. (1992), "Mediators in the architecture of future computing systems," *IEEE Computer*, vol. 25, no. 3, pp. 38-49, March.

Wiederhold, G. and Genesereth, M. (1995), "The basis for mediation," in Laufmann, S., Spacciapieta, S., and Yokoi, T. (eds.), *Proceedings of the 3rd Conference on Cooperative Information Systems (CoopIS-95)*, Vienna, Austria, May.

Wooldridge, M.J. and Jennings, N.R. (1995), "Intelligent agents: theory and practice," *The Knowledge Engineering Review*, vol. 10, no. 2, pp. 115-152.

Index

List of Contributors

Martin Beer
School of Computing and Management Sciences
Sheffield Hallam University, City Campus
Howard Street, Sheffield S1 1WB
U.K.
m.beer@shu.ac.uk

Charles Branas
Center for Clinical Epidemiology and BioStatistics
School of Medicine
Univ. of Pennsylvania
Philadelphia, PA 19104-6021
U.S.A.

H.-J. Bürckert
DFKI German Research Center for Artificial Intelligence
Stuhlsatzenhausweg 3
D-66123 Saarbrücken
Germany
hjb@dfki.de

Gerd Doeben-Henisch
Knowbotic Systems GmbH + Co KG
Daimlerstrasse 32
D-60314 Frankfurt am Main
Germany
doeben-henisch@knowbotic-systems.com

Louwrence Erasmus
Knowbotic Systems GmbH + Co KG
Daimlerstrasse 32
D-60314 Frankfurt am Main
Germany
erasmus@knowbotic-systems.com

Stan Franklin
Institute for Intelligent Systems *and*
Department of Mathematical Sciences
The University of Memphis
Memphis, TN 38152
U.S.A.
stan.franklin@memphis.edu
www.msci.memphis.edu/~franklin

P. Funk
DFKI German Research Center for Artificial Intelligence
Stuhlsatzenhausweg 3
D-66123 Saarbrücken
Germany
funk@dfki.de

A. Gerber
DFKI German Research Center for Artificial Intelligence
Stuhlsatzenhausweg 3
D-66123 Saarbrücken
Germany
agerber@dfki.de

Joachim Hasebrook
Educational financial portal [efiport] AG
Sonnemannstr. 9-11
D-60314 Frankfurt am Main
Germany
jh@efiport.de

John Holmes
Leonard Davis Institute for Health Economics
and
Center for Clinical Epidemiology and BioStatistics
School of Medicine
Univ. of Pennsylvania
Philadelphia, PA 19104-6021
U.S.A.

Wei Huang
Department of Computer Science
The University of Liverpool
Peach Street
Liverpool, L69 7ZF
U.K.
huang@csc.liv.ac.uk

Doug Ivins
Center for Clinical Epidemiology and BioStatistics
School of Medicine
Univ. of Pennsylvania
Philadelphia, PA 19104-6021
U.S.A.

Catholijn M. Jonker
Department of Artificial Intelligence
Vrije Universiteit Amsterdam
De Boelelaan 1081a
1081 HV Amsterdam
The Netherlands
jonker@cs.vu.nl

Stephen Kimmel
Leonard Davis Institute for Health Economics
and
Center for Clinical Epidemiology and BioStatistics
School of Medicine
Univ. of Pennsylvania
Philadelphia, PA 19104-6021
U.S.A.

M. Klusch
DFKI German Research Center for Artificial Intelligence
Stuhlsatzenhausweg 3
D-66123 Saarbrücken
Germany
klusch@dfki.de

Jiming Liu
Department of Computer Science
Hong Kong Baptist University
Kowloon Tong
Hong Kong
jiming@comp.hkbu.edu.hk

Hanh H. Pham
Department of Computer Science
State University of New York at New Paltz
75 S. Manheim Blvd. Suite 6
New Paltz, NY 12561
U.S.A.
pham@cs.newpaltz.edu

C. Russ
DFKI German Research Center for Artificial Intelligence
Stuhlsatzenhausweg 3
D-66123 Saarbrücken
Germany
russ@dfki.de

Hüseyin Sevay
Department of Electrical Engineering and Computer Science
Information and Telecommunication Technology Center
University of Kansas
Lawrence, KS 66045
U.S.A.
hsevay@ittc.ukans.edu

Barry G. Silverman
Leonard Davis Institute for Health Economics
and
Systems Engineering Department
University of Pennsylvania
Philadelphia, PA 19104-6315
Barryg@seas.upenn.edu
http://www.seas.upenn.edu/~barryg/index.html

Andrew Sixsmith
Department of Primary Care
University of Liverpool
P.O. Box 147
Liverpool, L69 3BX
U.K.
sixsmith@liv.ac.uk

Jan Treur
Department of Artificial Intelligence
Vrije Universiteit Amsterdam
De Boelelaan 1081a
1081 HV Amsterdam
The Netherlands
treur@cs.vu.nl

Costas Tsatsoulis
Department of Electrical Engineering and Computer Science
Information and Telecommunication Technology Center
University of Kansas
Lawrence, KS 66045
U.S.A.
tsatsoul@ittc.ukans.edu

Ransom Weaver
Systems Engineering Department
University of Pennsylvania
Philadelphia, PA 19104-6315
U.S.A.

Jianbing Wu
Intelligent Systems Group
University of Calgary
Alberta
Canada
jbwu@enme.ucalgary.ca

Ning Zhong
Department of Information Engineering
Maebashi Institute of Technology
460-1 Kamisadori-Cho
Maebashi-City, 371-0816
Japan
zhong@maebashi-it.ac.jp

MIX
Papier aus verantwortungsvollen Quellen
Paper from responsible sources
FSC® C105338

If you have any concerns about our products,
you can contact us on
ProductSafety@springernature.com

In case Publisher is established outside the EU,
the EU authorized representative is:
Springer Nature Customer Service Center GmbH
Europaplatz 3, 69115 Heidelberg, Germany

Printed by Libri Plureos GmbH
in Hamburg, Germany